Nature in the Round

NATURE IN THE ROUND

A Guide to Environmental Science

Edited by

Nigel Calder

The Viking Press / New York

CONTENTS

CONTENTS

CONTENTS

ACKNOWLEDGEMENTS

Acknowledgement is due to the following for illustrations (the numbers refer to the pages on which the illustrations occur):

26: W. H. Freeman and Co. for permission to use a diagram from *Resources and Man*, M. King Hubbert, published for the National Academy of Sciences;

84: Methuen and Co. Ltd for permission to use a map from *The Changing Map of Asia*, ed. W. Gordon East, O. H. K. Spate and C. A. Fisher;

113: The University of California, Berkeley, for permission to use a diagram from a report by L. E. Anderson;

248-9: André Deutsch Ltd for permission to adapt two diagrams from *Only One Earth*, B. Ward and R. Dubos.

'*Scientific research and development in the context of environmental problems . . . must be promoted in all countries. . . .*'

United Nations Declaration on the Human Environment, 1972

'*. . . for confronting vested interests only the best research will do.*'

Ray Gambell, this volume

PART I
INPUTS OF
EXPERTISE

Editor's introduction to Part I

'Environmental science does not exist as such,' a noted environmental scientist tells me firmly. If he is right, then this book is offered as a step towards its invention. The analogy is with medical science, which draws eclectically upon a range of other sciences for anything that may help the patient. In our case, the patient is the wonderful, vulnerable web of living and non-living processes on and near the surface of the Earth. But we are still in the heyday of quacks and nostrums and nobody really knows whether the patient is at death's door or just off-colour.

A definition: environmental science is any research that may assist in prudent management of the global and local environments of life on Earth. Despite its catch-all character, the subject is not shapeless. A unity is supplied by nature, in the real systems investigated, while the emphasis on management focusses attention on the frontier between the human and non-human domains. The nature that we want to see 'in the round' certainly includes man, a product of nature.

In this book expert contributors discuss, with real-life examples, many different approaches to the environment. In the first part, they show how each of various classical fields of learning can now contribute to environmental investigations. The second part cuts the cake another way, with greater emphasis on common methodologies for environmental science. Although this compendium is not conceived as an all-embracing textbook, the topics discussed span something like a continuum from geochemistry to urban planning.

To emphasize the latent unity of environmental science and its essential practicality, Part I opens with an account (Chapter 1) of an unusually comprehensive study of a human ecosystem around a lake in Malawi. Chapter 2 reflects on environmental biology and the ecological tradition that provides a ready-made theme for environmental science. But our subject is 'ecology plus', and the most obvious extensions are into the soundly established fields of

3

the earth scientist (Chapter 3), the meteorologist (Chapter 4) and the agricultural scientist (Chapter 5).

Agriculture is a bridge, as we then begin to look more particularly at man and his works. The anthropologist (Chapter 6) can tell us about the ecology of the hunting way of life for which our species originally evolved. Our ancestors exchanged the bow for the plough, and archaeology and environmental history (Chapter 7) show how far-reaching have been the consequences. For the more recent effects of industrialization, the urban geographer (Chapter 8) takes the condition of Japan as an object-lesson.

Our modern urban areas could be arranged much more intelligently, as the chapter on physical planning (Chapter 9) makes clear. Medical science though (Chapter 10) warns us of the ever-present dangers of disease wherever large numbers of human beings are gathered and it emphasizes the quality of the total environment as a determinant of human health. To guard that quality, a special obligation falls on the engineer (Chapter 11) to consider how his technology interacts with the environmental system, while economics (Chapter 12) helps to show us more precisely what the options are, as between enjoying and conserving our available resources. As explained in the last contribution in Part I (Chapter 13), the new tendency in geographical theory is to exchange maps for computer models – a methodology to which we return in Part II.

If the first contribution, which begins opposite, differs in character from more globe-embracing chapters that follow it, that is deliberate. Our aim throughout the book is to temper generalizations with down-to-earth instances, for the sake of both rigour and clarity. How better to start than with fishermen of Malawi, who are out to win, from their complex environment, a life worth living?

Nigel Calder

One human ecosystem   I

A LAKE IN TROPICAL AFRICA
Margaret Kalk

Lake Chilwa in Southern Malawi has been a backwater of a tropical developing country. Geology and history left the lake, and the well-peopled plain surrounding it, as a self-contained ecosystem (see map overleaf). The welfare of its human populations, and of the fish and crops on which they depend for income as well as sub-sistence, is dominated by a peculiarity of the local environment. The lake is shallow and has no outlet; its level varies from year to year, according to the pattern of rainfall. In years of high level, the lake provides a relatively good living for the people of the plain, and the fish is also a major source of protein for other Africans in the incipient industrial areas of Southern Malawi. When the lake is low, people for 100 kilometres around go short of fish, and meat is too expensive to be a substitute. The demand for fish has always been greater than the supply.

For twenty centuries Lake Chilwa was at a cross-roads of Africa. People migrating from both east and west to settle in the fertile plain brought Iron-Age skills, crops and rural technology from far and wide. Unhappily it lay on a route to the sea used by slave-traders. The missionaries and the colonialists, after the mid-nineteenth century, opened up river and lake travel, and built a road and a railway from south to north which altered the communications

Since 1966 Margaret Kalk has been director of the Lake Chilwa Co-ordinated Research Project of the University of Malawi and a professor in the university's Department of Biology. British by origin, she has previously taught at the Universities of the Witwatersrand and Melbourne, specializing in physiological ecology.

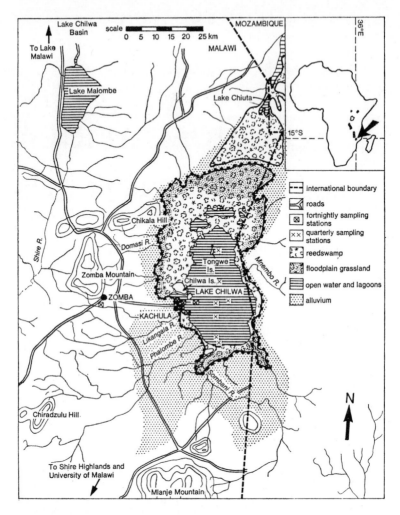

Lake Chilwa lies at an altitude of 620 metres above sea level. It has at present an area of open water of about 700 square kilometres, about 2 metres deep, surrounded by bulrush swamp of an equal area (broadest in the north) which is organically part of the lake. An additional 200 square kilometres of grassland is inundated in the wet season. The agricultural alluvial plains around it in Malawi are about 2,000 kilometres in area and support 300,000 people; the eastern shore abuts on Mozambique. The western plain has access to the town of Zomba 30 kilometres away. (Map adapted from a compilation by C. Howard Williams)

axis of Malawi (then Nyasaland) and left the Lake Chilwa Basin isolated. Thereafter the area settled into a simple integrated economy of farming, fishing, hunting and food-gathering, organized in matrilineal societies.

The modern phase began in 1958. It is dated by the technological innovation of nylon nets, replacing the traditional nets handwoven from local fibre. Since then the economy has been changing from near-subsistence to a cash economy, still scattered but of quite high potential in both fisheries and agriculture. It may also have a future in tourism and minor exploitation of rare-earth minerals.

While the fishermen of Lake Chilwa began landing record catches with their nylon nets, the University of Malawi was being founded only 100 kilometres away. The scientists and social scientists of the new university were first struck by the emerging economic importance of the fisheries. As we looked more closely, we saw that the development of Lake Chilwa and its environs displayed, in miniature, many of the issues that environmental science has to face.

There were the universal problems manifest in a minor way: conflicting interests in the use of land and water; difficulties in transport and communication; inadequate planning in technology; lack of cultural amenities in a densely populated area; destruction of the environment (by fire); dangers of pollution from insecticides and fertilizers; and so on. Besides these there were the problems of a newly developing country, notably health hazards, new ways of rural life conflicting with tradition, and delay in the provision of the educational, medical, social and commercial services which are part of a cash economy. On the positive side there was no hunger, no competition, no industrialization and its consequent evils, no overcrowding.

The investigation

By 1965 Lake Chilwa was yielding as much fish as was Lake Malawi, which is ten times larger but poorer in nutrients because of its immense depth. The methods of fishing on Lake Chilwa ranged widely: valved basket traps in the streams, long-lines with hundreds of hooks, seine nets drawn straight on to the few sandy beaches, stout dug-out canoes, driven by bamboo poles, from which shallow gill-nets were set overnight, and large basket scoops held in front of canoes for catching small fish. There were also a very few plank boats driven by outboard engines. The fish catch fell in 1966, as our studies were beginning, and word came that the sizes of the fish were smaller. As a remedy the mesh-size of the nets was increased by a government propaganda campaign.

The alluvial plain up to the foothills of the catchment area was used diversely for growing maize, cassava, rice, millet and sorghum, cotton and tobacco, with widely different success in different parts. There was only one wide dirt road from the largest western landing beach to the town at Zomba and only three dirt tracks, impassable to anything but bicycles in the wet season, to serve the south and north. There was no electricity nor any telephones in the area.

Several pre-existing reports suggested other problems. Red locusts had periodically devastated the area until 1948 and the most recent large-scale spraying by helicopter had been done as recently as 1962 by the International Red Locust Service of Zambia. The northern part of the Chilwa Plain was suspected of being a potential outbreak area; that the red locusts bred there was evident, because they appeared on the Zomba market as food. Leprosy, malaria, bilharzia and hookworm seemed prevalent and, while a large-scale leprosy campaign had been initiated, taking medical help into every home in the villages, no mosquito survey had been made. Snails capable of transmitting bilharzia certainly occurred in the mouths of streams and nearby marshes, but it was not clear whether they were actually infected by human parasites. One wondered, too, about a naturally radioactive plain on Chilwa Island and its possible ill-effects on the people who grew maize there.

Lake Chilwa was known to be a relic of a Pleistocene lake about four times larger, because the Geological Survey had found wave-cut terraces up to 33 metres high on the hills around and alluvial silt to depths of 55 metres in the lake. The rate of encroachment of the swamp into the lake was not known, but was suspected of being rapid. The future of the lake was uncertain. The plain had been teeming with game (elephant, zebra, antelope and buck) a hundred years ago; today only a few bushbuck, reedbuck and rodents are hunted but the grassland supports about thirty-five thousand head of cattle.

Many possibilities for interdisciplinary environmental research thus offered themselves, and a co-ordinated research project was set up. Initially, we at the university hoped to offer predictions of future lake levels, to seek a scientific basis for better management of water and soils and to suggest locations for amenities for the people.

In 1968 every problem was accentuated dramatically by the temporary disappearance of Lake Chilwa. The whole lake bed dried up for a few weeks at the end of the dry season. The rains of 1968-9 then brought the depth of water back to two metres, only a little less than peak levels of the 1950s. The water, though, was very turbid and quite unlike the shrinking lake of 1966-7 in its chemical and physical properties. The scientific and practical challenge of this event was a great stimulus to our research and our efforts have broadened and intensified, as summarized overleaf.

The future of the lake

On the plains are the permanent villages of fishermen who in most cases are also small farmers. Family farm holdings may vary from less than half to more than one hectare, and are usually of mixed type. Villages cluster around sources of fresh water, for many wells are brackish. The houses are built of mud, plastered over a framework of branches that are neatly tied together and covered with a thatched roof with an overhanging canopy all around for shade. At the lakeside, fishermen build temporary reed huts where they may live for a few days or weeks up to 10 kilometres from home, while their wives and families, and possibly employees, continue farming. In the northern swamp some fishermen site their huts on sudd – floating mats of vegetation – and keep their catch alive and fed in large baskets under water, until they make their weekly trek to meet the traders. Others build their huts on poles in the shallow lake edges, whence gill-nets can be set and watched all night. At the few sandy beaches 'sailors' villages' have grown up with small grocery shops, canteens and bars. So there is a migrant element in society, with the usual shortcomings.

Fishing is pursued fairly regularly during the eight drier months of the year and is considered the best and quickest source of income for paying taxes, for providing school fees, for purchase of beds, lamps, bicycles, sewing-machines, clothes, and for improving farming. Primary education among fishermen and the villagers generally is lower than the 35 per cent attendance average in Malawi. Secondary schools exist only on the periphery of the area near the towns. Distances from hospitals or dispensaries are also greater than the average in the country as a whole, many parts of the plains being 24 kilometres from the nearest clinic in 1966. But the system

9

Lake Chilwa Co-ordinated Research Project (University of Malawi)

Topics investigated 1966–72	Some practical conclusions
physical geography and lake history; climatic patterns; inflow to lake versus evaporation	changes in lake level understood; long-term drops in lake level remain a problem; water impoundments or diversion under investigation
water analyses – physical and chemical (lake, swamps, rivers, boreholes, salt springs)	'new' open lake is less salt and more turbid than 'old'; brackish wells a problem
species and biomasses in the lake (bacteria, algae, zooplankton, bulrush, bottom animals, fishes)	rotting swamp is a vital source of nutrients; recolonization of lake by fishes from reserve in lagoons remarkably rapid; their diet changes according to food available
swamp encroachment (?) and ecological effects	burning should be controlled; growth of swamp plants is the major primary production for the lake
population dynamics of economic fishes	new methods of fishing and research into fishing potential
birds as fish catchers	minor importance
red locusts: breeding habits, predators	locusts a possible threat; predators numerous
succession of grass species near lake	relevant to locust behaviour
interrelationship of human occupations (fishing, fish-trading, farming, cattle-keeping)	fishermen or fish-traders are farmers who put savings from fish enterprises into farming cash-crops
economic and welfare surveys of fishermen (during decline and after recovery of lake)	fish catch is by far the most important factor in prosperity, but cultural amenities are needed
role of fish-traders who process fish	division of labour and mode of distribution work well

response to major rice-development (land reassignment, irrigation, machinery)	radical innovations are accepted if managed well
insecticides in irrigated areas	use suspended while biology of pests investigated
combined rice and vegetable farming	promising, but fresh water and transport to market are problems
snail vectors of bilharzia in the lake, streams and swamp (no snail vectors of cattle liver fluke)	probably no human bilharzia in the lake; but disease is prevalent in the wet season from snails in streams; Chilwa cattle are free from liver fluke
mosquitoes and malaria on plains	serious and worsened by paddy rice-growing
natural radioactivity of Chilwa Island	
(a) irradiation of inhabitants	slight – detectable only in women farmers at end of dry season
(b) uptake in crops and fish	still under investigation
potential mining of rare-earth minerals	roads unsuitable at present
potential for tourism	lack of shade, lake turbidity, apparent poverty of temporary villages, lack of amenities are deterrents
transport and communication	wet-season motor roads badly needed; new roads are under construction
regional planning	schools, clinics, shopping and community centres needed

The project involves seven university departments: biology, chemistry, physics, mathematics, geography, economics and sociology. About thirty members of staff have contributed to the work over the years. Financial support has come chiefly from the Leverhulme Trust Fund, the British and Malawian Governments and the International Biological Programme (UK). The work has depended on the good will and help of villagers, chiefs, headmen, the Government of Malawi and Congress Party officials.

of 'self-help' for building schools and the extension of clinics for the under-fives are proceeding well, so that the 1966 census figures are already out of date.

The course of development of the area depends on the success of fishing, the present prime source of income, and hence on the stability or rate of recession of the lake in the near future. A million years ago the lake was four times its present size, its recession having provided the present fertile plains. Several geological events have affected the size of the lake: at least one river was diverted to the Indian Ocean; ground movements in the Later Stone Age tilted the lake bed to the east and made the western side shallower; a sand bar 16 metres high sealed off a former outlet of Lake Chilwa draining to the Indian Ocean; and lastly there is the continuous deposit of large amounts of silt and clay brought by the rivers to the lake bed every year.

From oral testimony we have learnt that the depth of the lake considered 'normal' varies from 2 to 5 metres. Recessions severe enough to interfere with open-water fishing occurred in 1879, 1922 and 1968; less severe recessions expose the periphery of the lake bed at irregular intervals of fifteen to thirty years. The phenomenon appears to depend on the two patterns of rainfall on the southern and western catchment areas, which vary greatly from year to year, though independently of one another. If there should be a coincidence of low rainfall in the two drainage areas for three years or more then more or less severe recessions occur. The usual seasonal rise and fall of the lake level was about 0.75 metre in the 1950s, but from 1964 to 1968 evaporation exceeded rainfall and drainage every year, until the open lake dried out. Water still remained in the lagoons and a few sandblocked permanent streams; in these parts fishing continued, and perhaps even overfishing, considering the necessity for restocking the 'new' lake.

Nevertheless, within three years of the catastrophe of 1968 fish catches were back to half the record yields of 1965. The continued study of the lake and improved management of fisheries *is* therefore worth while. It has already become apparent that, far from detracting from the productivity of the lake, the large area of swamp surrounding it contributes a major supplement to the food chain. Annual bacterial decay of its lush growth compensates for want of productivity of the lake algae, which is fairly low except for the few weeks of surface blooms of blue-green algae that occur before the flood waters arrive each year. The most encouraging sign for the future is that the 'new' lake is much less salt than the old, the clay in suspension and the soft bottom having absorbed the salts.

The technological advantage of an all-weather road, greatly improved in 1969, with organized transport for fish traders, is the key to present success. A large fleet of trucks carries the fish-traders, together with their bicycles and 50 kilograms of fish in each of their baskets, to the two urban markets of Zomba and Blantyre, or else drops them off to cycle to rural markets along the way. It has proved to be a very economical and efficient method of distribution. The separation of the labour of fish-processing and marketing from fishing enables many more people to derive their livelihood from the lake.

The Fisheries Extension Service of the Malawi Government is active in promoting improvements in fishing and kiln processing of fish that are smoked and dried. *Tilapia* (a cichlid) is mostly sold fresh but may be smoked, *Barbus* (a minnow) is sun-dried on mats, while *Clarias* (a catfish) is gutted and kippered. A new method of paired-boat trawling has been proved to be far more efficient than traditional methods of fishing. A boat-building yard is now making small power-craft; the nets are constructed by adding floats and weights of local materials to the nylon netting manufactured in Blantyre. Credit is extended to purchasers of the new gear, who are as yet mainly Young Pioneers of the Malawi Congress Party, trained in its use. Thus fish of the 'open water', dangerous for canoes, will also be tapped.

A local 'theory' that the production of fish is inexhaustible is, of course, being carefully investigated. There are no obligatory predators among the fish and the pelicans and cormorants are few and near the swamp. The diets of the economic fish turn out to be various and optional. The food chain begins with the wealth of the annual growth and decay of the bulrush swamp and its micro-fauna and micro-flora. *Clarias* and *Barbus* breed in the swamps, where the water is clear and the production of zooplankton (small animals) is high. Zooplankton concentration is only moderate in open waters, but the daily mixing and turnover of the shallow waters provides for the growth of phytoplankton (small plants) which is being measured. The only direct consequence of the new turbidity appears to be the complete disappearance of chironomid fly larvae from the bottom water, except in the swamps and the lake fringe. But we have

13

still to study and calculate how many powered craft the lake can support to supplement the adaptable canoes, which can penetrate the swamp channels, and the traditional long-lines, shore nets and traps.

The long-term study of the bilharzia-bearing snails around the fish landing beaches has not yet revealed infection with the human parasite, although 8 per cent had cattle parasites, and some were hosts to a stage in the life cycle of a fluke which passes through snail, copepod, fish and bird. The last kind of parasite was probably responsible for the phenomenon of widespread blindness in *Tilapia* during the decline of the lake in 1967.

Agriculture and rural life

On the Phalombe Plain to the south-west of Lake Chilwa, sheer topography enforces a natural integration of fishing, fish-trading, farming and cattle-keeping. This balanced use of land and water must have operated for a hundred years or more. The traditional practice of growing paddy rice where the water table is high, cassava on mounds and maize and vegetables in the drier season remains the basis of subsistence.

Farmers who are not also fishermen or fish-traders are usually the owners of cattle which live on free land, rich in fodder the whole year round. The government has built cattle dips at accessible spots and small supplies of meat are sent to rural and urban markets. The chief incentive to change to a cash-crop economy has been provided by access to the tea plantations and town on the Shire Highlands. Income from fishing and fish-trading has been used to boost cotton farming. In this area is a depot where cotton is graded and bought by the Agricultural Development and Marketing Corporation. The weak link which prevents a more rapid transition to a stable consumer society is the road to the Shire Highlands, which is impassable to motor vehicles in the wet season. There is a large health centre but this potentially dynamic area also lacks a regional centre where educational, social and commercial services would help realize a higher standard of living.

A very different farming enterprise is the rice development project along the banks of another river on the west of the lake. Here the regime has not followed traditional lines but is highly organized by Taiwanese. Land tenure now depends on successful use of land and not on hereditary matrilineage. The land is tilled by tractors and divided into plots of 0.15 hectare allocated by officials and not by the Chief. Routine work is encouraged with a rigid timetable for preparation, planting, weeding and harvesting; irrigation from the fresh stream is carefully nurtured. Vegetables are also grown for sale, although transport to markets is difficult. The farmer has to follow the instructions of officials; the crops belong to him but he is expected to spend his profits on school fees, clothes, farm equipment and household requirements, not on beer, which was hitherto the traditional basis of co-operation between neighbours in farming and house-building.

The pilot project proved successful because it was joined by the Chief and his relatives, some Young Pioneers and fishermen with savings, because of the participation of the Taiwanese as workers and, above all, because of the monetary success of the first and successive years. The favourable response shows that traditional culture is not a force resistant to innovation. Special advantages have been the high density of population and the ethnic diversity (Lomwe, Yao, Man'anja and Nyanja) which had already changed social standards. The rice schemes in this pattern are growing in size and increasing their yields, in spite of an initial fear that irrigation would further deprive the lake of water enough to maintain fishing potential. In soil deficient in nitrogen, irrigation canals have the advantage of encouraging the preferential growth of blue-green algae, some of which can fix nitrogen from the air.

We are, however, concerned with hazards of irrigation. Bilharzia and hookworm are more easily spread, and malaria mosquitoes breed in the shallow water, whereas on the lake during the day there are few mosquitoes. Regrettably, mosquito-eating fish have not yet been introduced to reduce human suffering, nor are the necessary boots and latrines available. Cultivation of rice on a large scale also encourages infestation by pests. Uncontrolled use of insecticide has however been suspended, while the ecology of the rice pests is being studied. Infestation is not at present high.

The nearby Cotton Research Station tests out many insecticides on the plethora of imported cotton pests and we shall monitor the fate of the residues, hoping that absorption by the suspended clay of the lake may lessen the problem.

Around this lake in tropical Africa, then, environmental science is a going concern. Our aim is certainly not to impede economic growth but to assist it and safeguard it. While an interdisciplinary study such as ours can collect and interpret data and suggest the pointers to progress, the future of the people of the plains does not depend only on the natural wealth of the lake and soils nor on the efforts of the people themselves. Regional planning is vital, especially of transport, health centres, schools and those cultural and commercial amenities on which the cash incomes can be fruitfully spent for better living. This is now government policy and, if the main assets of fish and rice can be maintained, the future should see flourishing lakeside communities at Lake Chilwa.

References and further reading

S. Agnew, 'An indigenous growth towards a cash economy in Africa', *Geoforum*, 12 (1972).

S. Agnew and M. Stubbs (eds.), *Malawi in Maps* (London 1972).

A. Chilivumbo and P. Phipps, 'The fishermen of Lake Chilwa', *Journal of the Institute of Social Research of Malawi* (in preparation).

M. Kalk (ed.), *Decline and Recovery of a Lake* (19 research reports) (Zomba 1970).

B. Pachai (ed.), *The Early History of Malawi* (London 1972).

C. Ratcliffe, *Experimental Small Craft Pair Trawling, Lake Chilwa* (Zomba 1971).

THE ECOLOGICAL TRADITION

Kenneth Mellanby

For centuries country clergymen and other amateurs have observed wildlife and have kept careful records of the distribution and behaviour of many species of animals and plants. Few have recorded these facts in a permanent form; fewer still have had the literary ability of Gilbert White or Izaak Walton, so that their work has been known only to a limited public. Until recently most naturalists were also inveterate collectors, and some of their herbaria or accumulations of bird eggs form useful sources of information about the state of the countryside in the years, sometimes as long ago as the eighteenth century, when the specimens were originally preserved. The successful collector's interests were wider than even he realized, for to learn to distinguish between a 'good' and a 'bad' area for collecting meant, in effect, learning to pick out the important factors in a species' habitat.

Fishermen and hunters have been primarily interested in the particular species which have given them sport, but both the sportsmen themselves and the gamekeepers and others whom they have employed have often come to realize that even a 'preserved' game species lives in an environment in which many other factors, living and non-living, play essential parts. Few naturalists, sportsmen or

Kenneth Mellanby has been director of the Monks Wood Experimental Station of the British Government's Nature Conservancy since its establishment in 1961. He is an entomologist by profession, and edits the journal *Environmental Pollution*. He was the first principal of University College, Ibadan, Nigeria, from 1947 to 1953.

gamekeepers look upon themselves as ecologists, yet their success depends on what we can now clearly recognize as sound ecological observation.

What is ecology?

There are many definitions of ecology, but these all boil down to some expression about the study of the interrelationships between living organisms and their environment. From what I have already said, it must be clear that I do not think that there is any hard and fast division between 'natural history' and 'ecology'; indeed the term 'scientific natural history' would be a satisfactory definition of ecology. However, while I should always expect a good ecologist to be a naturalist, it must be admitted that many who are deeply interested in some phase of natural history do not possess the qualities or admit the disciplines required to merit the name ecologist. In passing I would like to deplore the common assumption of the name today by many, particularly those in the communications industry, who seem to me to have no right to call themselves ecologists. A concern for the environment, no matter how eloquently expressed, is no substitute for the hard work and careful observation necessary to the training of the true ecologist.

Though many naturalists have, over several hundreds of years, practised what we now call ecology, the emergence of the subject as a scientific discipline is comparatively new. Charles Darwin himself was a sensitive observer of interrelationships in living nature, and gave all biology its present theoretical base, but stopped short of modern ecological concepts. Haeckel, Darwin's contemporary, spoke of 'nature's household', a useful expression of ecological principles. The transformation has occurred in the twentieth century and, in retrospect, the efforts of the late Sir Arthur Tansley in the botanical field and of Charles Elton in the zoological are plainly of paramount importance. Nevertheless we owe almost as much to a host of workers in all parts of the world, of whom E.P.Odum in America and J.Braun-Blanquet in continental Europe may be cited as examples.

The rapid progress of ecology has depended first on the genius of these leaders and secondly on the traditional interest in natural history which allowed the relevance of their ideas to be realized. Points where progress has been notable include the study of the effects of edaphic factors (those caused by the soil and the non-living substratum) and climatic factors on individual organisms, and on plant and animal associations. There has also been an increasing reliance on the proper statistical treatment of numerical data, and on studies of the energy budgets of plants and animals and selected areas that they occupy. Ecology is becoming 'quantitative natural history' to an increasing extent.

Once the principles of scientific ecology had been enunciated, by the mid-1930s, they gave a much-needed direction to the fieldwork of professional scientists and amateur naturalists alike. Ecology depends on observation, but equally important are the recording and analysis of the data observed and the storage of data and analyses in such a way that they are easily accessible to others. That may seem trite – would that it were! It is a tragedy that the majority of the records of the older naturalists have either been lost or remain buried in some form that makes retrieval difficult or even impossible. Many newer records are equally inaccessible, representing count-less man-years of fieldwork that benefit no one but the observers themselves.

Within the past twenty years there has been some progress in improving this situation. For example, we had the plant-mapping scheme of the Botanical Society of the British Isles. Here, compara-tively sophisticated techniques were used to record and analyse the distribution of all British flowering plants, drawing when possible on old information from many sources. The result was a set of detailed maps of the distribution of the various species and, in some cases, maps showing the changes of sites occupied by rare and possibly endangered species. This pilot scheme encouraged the establishment of a Biological Records Centre at Monks Wood Experimental Station of the British Nature Conservancy. The centre uses the most modern methods to cope with information about an increasingly large number of species, including birds, mammals, amphibia, many insects and various non-flowering plants.

This work has catalysed similar developments in many other countries. Schemes such as the European flora maps and a whole series of data collections made for the International Biological Programme are fostering international collaboration. We are at last beginning to obtain a baseline for future studies of many species. This gift to our scientific successors makes us even more aware of our own present disabilities. The lack of numerical information about earlier days makes it difficult to quantify the effects of pollution and other current environmental changes.

The ecosystem

Ecologists study 'ecosystems', now a familiar term, but also one which is differently interpreted by different scientists. In general it is used for an area with a distinct community of plants and animals and takes into account the interrelations among the living organisms and between them and the non-living environment. Thus we may speak of a 'woodland ecosystem', and this is a useful concept even though it has its limitations. In theory an ecosystem should be a self-contained unit but, unless we consider the whole of our universe as a single ecosystem, pedantry on that score is pointless. We are coming increasingly to realize that there is 'Only One Earth' (the slogan of the United Nations environmental conference of 1972) and that all its parts are interrelated. As life on Earth depends for its energy on the Sun, even our planet is not a closed system.

In practice it is still convenient to divide the world into a vast number of ecosystems each of which, individually defined, can serve as a focus for particular studies. The divisions are not hard and fast. An ornithologist, for instance, might treat a stretch of woodland as a single unit, even though he knows that many woodland birds depend partly on surrounding farmland and that the woods may provide temporary refuges for species that normally live elsewhere. Then an entomologist may come along and dedicate himself to a single dead tree; this for him is also an ecosystem. The important notion is that, in each ecosystem, all possible factors, living and non-living, must be taken into consideration.

There is no such thing as a rigorous or static 'balance of nature'. The issue is the reverse one: how, despite continual change in every living community, does it still retain a certain basic stability? Ecologists have first to determine the reality of the various communities (or ecosystems or habitats) and to discover what plants and animals exist in them. They then have to try to discover the reasons why these organisms live in a certain place, and in association with other, characteristic, organisms. Some plants, for instance, flourish in limey soils and the insects that live on these plants are restricted to the same areas. Other organisms tolerate a wide range of conditions of soil and climate, and are therefore widely distributed.

Even within their favoured areas some plants or animals are common, others are rare. Some species remain almost constant in numbers year after year; some gradually become rarer or commoner; some exhibit, even in areas little affected by man, enormous fluctuations in numbers. The ecologist has often done well in determining *what* is happening in his chosen field but, in discovering *why*, he has only begun to make a little progress so far. Masses of data have been collected and submitted to a variety of sophisticated mathematical processes, including analysis by computer. No doubt this type of work will eventually prove of great practical importance, but that day still seems a long way off.

Ecologists and pollution

Nevertheless we do know a good deal about our flora and fauna and we can sometimes recognize when man's activities have drastic effects on particular ecosystems. One obvious instance is fresh water. Studies of streams, rivers and ponds have enabled us to determine which animals and plants live only in clean waters, and which are able to survive various levels of man-made pollution. The original work on which these findings are based was done by many scientists and naturalists over many years, but their conclusions can be applied in practice by the non-expert. Thus we have recently found that, when supplied with a simple kit, quite young children, even below the age of ten years, can quickly learn to find and identify a series of species of invertebrate animals (insects, crustaceans, worms) and relate them to the pollution levels of fresh waters. These children are able to recognize the *biological* effects of pollution, and even to recognize quite small sources which can be best detected by their effects on the balance of animal species in the locality.

This is just one of the instances where ecological methods can be used as a delicate – and simple – method of studying environmental pollution. It measures a real change, and is not entirely based on chemical analyses which may detect potentially dangerous substances at levels where they are probably harmless. As analytical tools, living organisms have the advantage that they do something to integrate the effects, over a period, of varying levels of pollution.

Periodical chemical analyses, however frequent, may miss a dangerous discharge of short duration, but which causes serious damage to
the environment.

Pesticides in the environment

Ecologists have made a particularly important contribution to the
understanding of the effects of pesticides in our general environment, and to establishing a scientific basis for controlling their use.
The only difficulty here has been our lack of basic knowledge. When
a member of the public says that some species of butterfly, for
instance, is becoming increasingly rare, and alleges that this is because of the use of insecticides, we realize our ignorance. First, we
may have no records showing whether the insects were in fact commoner some years ago. Secondly, even if we have this information,
the changes in numbers may have been caused, for instance, by the
weather (few butterflies are seen during a cold, wet summer), or by
changes in the habitat (such as ploughing up old pasture with many
food plants so that breeding is restricted). It is rarely self-evident
that the blame lies with high mortality caused by poisons applied by
man. These difficulties are reasons why many laymen find scientists
infuriatingly diffident about making dogmatic statements about
these topics, and why ecological studies of these problems are so
time-consuming.

Wild birds have turned out to be among the most delicate and
rewarding environmental indicators. For many years after their
introduction during the Second World War DDT and other organochlorine insecticides were accepted uncritically as a complete boon
to mankind. They certainly saved millions from insect-borne
diseases, including malaria and typhus, and they made a substantial
contribution to increasing world food supplies and so to feeding the
ever-growing human population. It was only when unexpected and
unwanted side-effects on wild birds were observed that the possible
disadvantages of these pesticides were realized.

Fortunately, and only because birdwatching has been so popular
for so long, we did have accurate data, going back many years, for
such predatory birds as the peregrine and the golden eagle. We
benefited from egg collections made during the last century – something our descendants may not have, because of our justifiable,
though scientifically restrictive, condemnation of egg-collecting

today. We found that these stable and persistent poisons were spread widely through the environment, by wildlife, in unexpected ways. When grain was treated with dieldrin – to protect it against the wheat bulb fly, a serious pest in many areas – the deaths of seed-eating birds were not entirely unexpected and the farmers who considered wood pigeons and other such birds as serious pests in their own right welcomed this outcome. What was surprising was the dramatic effect on hawks, particularly the peregrine, which were poisoned at one remove by eating contaminated pigeons and other seed-eaters. Peregrines were eliminated from many areas where they had bred successfully for centuries; even where the adults survived, breeding stopped. Golden eagles were contaminated from eating carrion, particularly sheep dipped in dieldrin; they tended to survive themselves but they, too, failed to breed. Comparisons of recent and pre-war eggs suggested that one reason was that sub-lethal doses of dieldrin and DDT caused the birds to produce eggs with thin shells, which were easily broken in the nest. The seed-eating birds, being rapid breeders, suffered less damage to the population though many individuals died.

These widespread effects were detected only because of ecological studies of wildlife. Parallel investigations in several countries and exchanges of information led to stringent restrictions in the more dangerous uses of persistent pesticides. Continuing ecological studies have revealed the success of these restrictions, which have brought about a fall in the levels of environmental contamination and a marked improvement in the breeding success of the birds most harmed.

These are examples of the practical value of ecological monitoring of our environment. Changes in the balance of animal populations may not always be easy to detect, but they may serve as early warnings of greater dangers to man and to the whole environment.

As the reader will have gathered, though, I take a modest view of the practical achievements of ecology so far. Having quite properly alerted governments and the public to general threats to the living environment, we usually lack the factual information needed, in any specific case, to give definitive assessments and advice. We must do what we can, relying on well-established principles and *ad hoc* research to bridge gaps in our data. We are particularly cautious, as we know that if the professional ecologist (as distinct from the 'ecological' publicist) is detected in any exaggeration or scientific mis-statement, his credibility, and possibly that of the whole profession, will be jeopardized. This could prevent him from continuing to play any useful role as an environmental watchdog in the future. As applied ecology ceases to be the Cinderella of biology, and as scientists from other fields begin to concern themselves more actively with environmental questions, we can look forward to important advances in knowledge and in the integration of knowledge. My only immodest claim would be that, however it develops in the decades ahead, environmental science will have to rely heavily on the ecological tradition, with its concern for the infinitely subtle interdependence of all life on Earth.

References and further reading

J. Braun-Blanquet, *Plant Sociology* (trans.) (New York 1932).
Charles Elton, *Animal Ecology* (London 1927).
Charles Elton, *The Pattern of Animal Communities* (London 1966).
Kenneth Mellanby, *Pesticides and Pollution* (London 1967).
E. P. Odum, *Fundamentals of Ecology* (Philadelphia 1959).
Arthur G. Tansley, *The British Isles and Their Vegetation* (Cambridge 1939).

THE
PHYSICAL BASE

B.M.Funnell

In many ways the earth sciences are central to environmental science. Not only have they provided the basis for man's mounting exploitation of minerals, fuel and other resources, but they also provide his principal guide to the possible effects of such exploitation on the global environment.

The earth sciences themselves are variously defined. They certainly include geology, geophysics and geochemistry, together with the related 'solid' earth sciences of geomorphology, sedimentology, palaeontology, petrology and geodesy. Most authorities would also include the 'fluid' earth sciences of oceanography, hydrology and meteorology. Together, these sciences study the physical and chemical characteristics of the Earth's interior, its surface, and the liquid and gaseous envelope which surrounds it.

Since the nineteenth century, specialists in all these fields have tried to view their respective problems globally, even though some major solutions on this basis have become generally available only since the mid-1960s. In this same very recent period, the concepts and technology of these planetary sciences have proved to be transferable to the study of the Moon, so that a comprehensive understanding of the lunar environment has been forthcoming in a very short time. As the earth sciences thus become comparative planetary

B.M.Funnell is a geologist and palaeontologist. He is currently dean of the School of Environmental Sciences in the University of East Anglia, Norwich, where he has been professor of environmental sciences since 1968. He is co-editor of *The Micropalaeontology of Oceans*.

sciences, they throw into relief the special qualities of the Earth that make it the only planet of the solar system suited to life as we know it.

Energy resources

Perhaps the most remarkable change in the human condition in the last few centuries, and particularly in the last few decades and years, is the rapid increase in the use of energy. Previously all the energy used by man in his various activities was derived more or less directly and instantaneously from the energy of the Sun, be it by the intermediary of photosynthesis and thence food or firewood, or by water and wind energy. Now we depend heavily on the fossil fuels – coal, petroleum and natural gas – which represent a chemical store of energy derived from sunlight that fell on the Earth's surface tens or hundreds of millions of years ago.

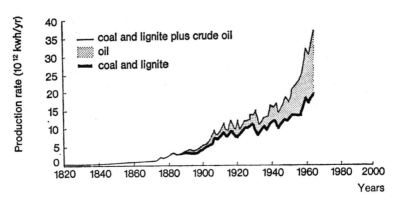

World production of thermal energy from coal and lignite plus crude oil (from *Resources and Man* 1969)

Finding the fuel resources that made possible the recent rapid increase in consumption shown in the diagram has been a remarkable success of the twin sciences of geology and geophysics. They have interpreted remotely, underground and undersea, the complex structures of rocks which contain oil, natural gas or coal. So successful has this search been, in fact, that we begin to perceive the absolute limits of all fossil fuels that are ever likely to be accessible at the surface of the planet. It is now possible to calculate the period of availability of such fuels given certain levels and rates of growth of consumption.

These estimates are no longer the guesswork they once were. Although different experts give somewhat different figures, some general statements can now be made more confidently and the governments and industries of the world should be expected to begin a process of adjustment. For example: the present rate of increase in demand for oil simply cannot be met for more than another twenty-five years or so; although coal reserves are much greater, they could not support high rates of economic growth for more than a hundred years or so. In both cases, if present rates of consumption, rather than rates of growth of consumption, were maintained, these resources would last significantly longer; perhaps fifty to one hundred years, and five hundred to a thousand years respectively.

Beyond these periods there will still be the continuing flux of solar energy to be captured by engineering works based on hydro-electric power, but these will be relatively small, and the contribution available from photo-electric and wind-driven sources can only be smaller still, compared with modern levels of power production. The input from planetary energy sources such as tidal and geothermal power will remain even smaller by comparison.

The only hope for continuing high levels of energy consumption lies in nuclear power. Here, too, the rapid geological evaluation of reserves of uranium and other heavy elements, since the advent of nuclear power, has already indicated limits to the reserves of uranium-235, the prime nuclear fuel. They are sufficient to support existing types of 'burner' reactors only for a comparatively short time. At present 'breeder' reactors are under development which, while producing power, convert uranium-238 and other heavy elements into fresh supplies of nuclear fuel. If that technology can be mastered the availability of nuclear fuels will be extended indefinitely. At that point the production of energy will become a problem not so much of finding fuel but rather of disposing of radioactive wastes. That, too, is a complicated matter involving geological considerations, because material buried at depth can find circuitous routes back to the surface.

In the meantime we may expect geological studies to become increasingly concerned with devising strategies for conserving fossil fuels as they become scarcer and more expensive.

Mineral resources

The exploitation of mineral resources for industrial purposes, as with fuel resources, is organized on a global scale. The same technologies and economic considerations which have led to the construction of supertankers for oil freight have led inexorably to bulk carriers for the long-distance transport of ores, in particular of iron ore.

Geologists long ago located the massive banded ironstone formations deposited during the Pre-Cambrian period, 3,200 to 1,800 million years ago. Their earlier explorations have only recently been vindicated by methods of mining and transport that make these ores available to the steel-makers of the world. Neither the black-band ironstones of the Palaeozoic Coal Measures (about 300 million years old) nor the minette ironstones of the Mesozoic rocks (about 150 million years old) are as amenable to large-scale mechanized excavation and processing as is the Pre-Cambrian ironstone. Although these younger ores were very important in the historical development of iron- and steel-making (and of geological theory) they are now being used less. For some other minerals, the situation is similar, and shortages of certain metals are forcing attention on extraction from ores previously considered too lean.

A distinction must be drawn between 'strategic' minerals which occur in high concentrations at relatively few places on the Earth's surface, and other minerals whose distribution is more widespread. Bulk rocks and minerals are often taken for granted, yet they present a particular problem. Although of low value in relation to volume, they may be required in very large quantities in engineering and building works. The requirements for aggregates, particularly for motorway foundations and for manufacturing concrete, may severely test the local availability of such materials in lowland regions – where urbanization and industrialization may be especially marked. It therefore becomes urgent to see that all potential sources of bulk minerals are geologically identified before they are sterilized by building upon them. With an eye to the future of the environment it is also important to assess the geological, geomorphological and hydrological context in which such resources are exploited; then the vast workings from which they are quarried can be programmed for restoration, for use in waste disposal, or for conversion into recreational lakes.

Water resources

Water, like energy, is essential for human activities. About 99 per cent of the Earth's water is saline and occurs in the oceans, 1 per cent is locked up in frozen form in polar ice caps and glaciers, 0.3 per cent is groundwater, hidden in the interstices of rocks underground, and only 0.01 per cent occurs in freshwater lakes and rivers where it is readily available to man.

As growing populations and expanding industrial production have increased the demand for water, earth scientists have been more and more involved in developing water resources, in bringing groundwater resources into supply and in increasing, by engineering works, the quantities of surface water impounded in artificial reservoirs. Lately it has even been proposed that the third major source of fresh water should be harnessed by bringing huge icebergs of polar ice by sea to water-hungry coastal states such as California.

Whilst at first sight it might seem that ground and surface waters are separate systems, this is not so. They exist in dynamic equilibrium. Both are fed by the same climatically-controlled input – rainfall – and in general withdrawals cannot exceed input except for limited periods. Nevertheless the hydraulic characteristics of ground and surface waters permit very different management policies. Water stored underground is out of sight, and usually it can be drawn upon without affecting conditions or amenity at the land surface. A groundwater reservoir can be emptied during dry months or years and in general, apart from certain irreversible conditions which must be guarded against, it will be refilled in subsequent wet periods. Groundwater can therefore be used to regulate the seasonal flow of surface rivers artificially, so that they maintain a constant flow suited to human requirements, rather than responding to seasonal variations in rainfall. Other positive attributes of groundwater are relative freedom from bacteriological contamination, and availability at considerable distances from surface supplies. For all these reasons the relatively difficult and intricate study of groundwater flow is currently attracting much attention from hydro-geologists, geophysicists and engineers.

Surface water impoundment is in many ways a less elegant method of storing water for human use. Compared with groundwater storage it is subject to higher rates of loss by evaporation and loss of storage capacity by siltation; it is more susceptible to biological contamination and there is also frequently loss of useful land and amenities when surface reservoirs are created. On the other hand the behaviour of water at the surface is better known and easier to observe, and the earth scientist's main role here is to locate and test suitable sites for their load-bearing and watertight properties.

Engineering works

The land surface supports all kinds of constructional work: roads, dams, housing, bridges, quarries, factories, harbours and so on. Using it safely requires a fundamental understanding of the way in which natural rocks, sediments and soils, as they occur in the landscape, can be loaded or cut away without detriment. This is an intricate problem: besides the inherent strength of rocks and their reaction to stress, their history as landforms, their present configuration and the pressure of any water that may be present in their interstices all help to determine the ability of natural surfaces to stand firm under structures loaded upon them.

The man-made structures themselves may be subjected to ephemeral stresses. In the earthquake belts that mark the boundaries between the great shifting plates of the Earth's outer shell, engineers collaborating with seismologists have evolved ways of making important buildings more or less earthquake-proof. Elsewhere, other questions are presented to the earth scientists. What is the risk of floods? From what direction will high winds come? How frequently and how high will wind speeds rise? To what intensity of storm waves will a harbour wall be subjected, and how high might the tide rise under exceptional conditions? The hydrologist, meteorologist and physical oceanographer will have to collaborate to answer such questions for the engineer's guidance.

Pollution in perspective

The whole surface of the globe is a process plant for dissolving salts and compounds from solid rocks, transporting them to the sea and accumulating them there as sediments or brine. The chemistry of natural waters affects their suitability for plant and animal life, and for human use. All the time the 'fresh' waters off the land are being titrated with the 'saline' water of the seas, in complex reactions in the estuaries.

The different environments produced by the various phases in this continuing process are congenial only to certain kinds of plants and animals. Introduce water, or rather its dissolved constituents, from another of these phases and you pollute the water for the original inhabitants. Introduce entirely foreign chemical compounds or excessive nutrients or alien microbial life, and you also pollute the water, and in the sense that most people understand the word. To the earth scientist, pollution is only a special case, at a particular time and place, of the type of chemical concentration or reaction that has characterized all past history at the surface of the planet. Such processes have concentrated sedimentary minerals and cycled the chemical compounds that have provided the raw materials for civilization. They have preserved chemical uniformity over shorter time-scales but have gradually altered conditions for life on Earth over many hundreds of millions of years.

Only the chemical oceanographer or geochemist can determine the baseline of 'natural pollution' against which to measure the impact of man. The incidence of short-term 'red tides' or eutrophic conditions in lakes, the long-term carbon-dioxide levels of atmosphere and ocean, even the mercury content of tuna fish – all these need to be seen in a geological time-perspective. What is detrimental for certain fish and human beings may be ideal for bacteria or dinoflagellates. And what would be lethal for modern plants and animals was essential for the origin of life!

Past environments

Much of the earth scientist's success has been due to the skill and imagination with which he has pieced together and reconstructed complex past environments from a wide variety of evidence. From these reconstructions he has been able to deduce the presence and disposition of rocks and resources. His habit of studying all aspects – physical and biological – of modern environments, in order to compare them with their ancient analogues, has in many ways pre-adapted and prepared the earth scientist for the current need to take a total view of the diversity of modern environments. It has also given him a vivid sense of how environments change with time.

Insight into the ways in which environments and conditions have altered in Quaternary time, approximately the last two million years, is particularly relevant to the human situation. The types of change which have characterized that period – dominated by advances and recessions of the polar ice caps – continue into the present time. In the study of the Quaternary period therefore, earth scientists are joined by botanists, zoologists, archaeologists, soil scientists and many other specialists, and the types of evidence available for interpretation in this period are more prolific than for earlier times. Nevertheless most of the principles by which such interpretation proceeds are those applied by geologists to much earlier periods and more remote happenings.

At the other end of the geologist's time-scale, some of the most interesting inferences concern the earliest periods of the Earth's history. For instance, some of the minerals in the banded ironstone formations, referred to earlier in this chapter, were laid down under water and yet they are chemically in a reduced state. This indicates that until the Earth was approximately half its present age the atmosphere and seas did not contain free oxygen; there is also other evidence to the same effect. During the period of the banded ironstones, oxygen produced by photosynthesizing bacteria and blue-green algae was directly taken up in oxidizing the iron dissolved in the water and precipitating it as ore. After a long period of more than a thousand million years this oxygen sink was filled; only then could free oxygen begin to accumulate in the atmosphere. Not until much later did this oxygen reach concentrations at which the great diversity of oxygen-dependent animals could exist. At present it is the balance between oxygen and carbon dioxide in the atmosphere which concerns us, regulated as it is by plant growth, the burning of fossil fuels, precipitations of limestones in deep and shallow seas, and other interrelated processes. Is the carbon dioxide increasing or decreasing, and against what past standard can it be measured? – these are questions currently requiring careful scrutiny.

The way ahead

From this sketch of some of the ways in which the earth sciences approach and perceive the environment it may be clear that these are strictly utilitarian and, at the same time, highly intellectual and curiosity-oriented. Much early development of the earth sciences arose from the commercial or economic needs of locating minerals, building canals, engineering public water supplies and navigating shallow seas. The modern counterparts of these operations are the location of oil and gas reserves, construction of dams and motorways, management of water resources and civil and military aviation. To the national geological and ordnance surveys, established in many parts of the world in the last century, are now added numerous government agencies concerned with hydrology, oceanography and meteorology.

On the other hand, such practical considerations also led to investigation, in the universities, of fundamental problems regarding the structure and evolution of the Earth. Stimulated by the atmosphere of Victorian science, including the great scientific explorations of the late nineteenth century and the acrimonious debates on evolution, very fundamental questions regarding the nature of the Earth were posed. Only recently, with the technology of the 1960s, have some of these problems been so excitingly resolved. Indeed, the theory of 'plate tectonics', which offers the first comprehensive explanation of major geological processes, is so new that its practical implications are only beginning to be worked out.

What, then, is the contribution of the earth sciences to environmental science to be at this juncture? Having provided such power for man to alter his environment, how can they be applied to conserve it? Their contribution is potentially most extensive. The earth sciences, like any other sciences, can be set to tackle any problems that the human spirit regards as important. Hitherto these have mainly been the acquisition of valuable materials for industry and profit, and the urge to understand the way in which the planet works. If the objectives are now to be those of conserving the character and resources of the Earth's surface, and of predicting the effect of man's activities on it, there will need to be some changes in priorities. Nevertheless I believe a combination of the practical and academic approaches will still be essential to achieve the best results.

The earth sciences in environmental problem-solving

The examples given here, of how the earth sciences can come together in the solution of particular environmental problems, are chosen almost at random. No doubt many readers can immediately think of others. However it is hoped they are representative.

33

THE EXPANSION OF THE CHICAGO URBAN AREA. Cities have always tended to grow, usually in a piecemeal fashion and sometimes with scant regard for geological and other environmental problems – which come to light only when damage has been done or opportunities missed. Planning for the present north-westward extension of the Chicago urban area was an example of a more systematic and sophisticated approach. It involved fieldwork by geologists specializing in superficial deposits, groundwater, and engineering and economic applications, together with laboratory studies by chemists, mineralogists, petrologists, stratigraphers and others.

Their investigations included surveys of land forms and the relation between agricultural soils and underlying geological strata. Mapping the distribution of sub-surface rocks and superficial sediments provided the basic information from which practical conclusions could be drawn about the occurrence of water-bearing strata, about proposals for water supply and waste disposal, and about sources of minerals and constructional materials. Determination of characteristics of near-surface rocks and sediments supplied essential data to the engineers. Results contributed by the earth scientists were combined with those of other experts in preparing the long-range plans which specified the optimum use of land during the development of the region, thereby avoiding premature building on sources of aggregate, the endangering of valuable water resources by contamination with the products of waste disposal, and so on.

THE IMPOUNDMENT OF THE WASH FOR FRESHWATER SUPPLIES. Providing enough water for the concentrated population in the drier, south-eastern part of Britain is a pressing task. One proposal is that of impounding fresh water behind embankments in the inner part of the Wash – the large, shallow bay on the east coast of England. Feasibility studies at present in progress require, as for Chicago, a variety of expertise. They involve geophysicists, engineering geologists, sedimentologists, hydrologists, physical oceanographers and chemical oceanographers in a large-scale interdisciplinary operation.

The detailed configuration of the seabed in the area is obviously of prime importance. Next, knowledge of the deposition of superficial sediments and deeper strata will allow an assessment of their likely reaction to the stresses of constructional work and eventual water pressures. The cost of the project will depend on the local availability of large quantities of constructional materials; its effectiveness on the quantity and quality of fresh water entering the dammed area from the rivers. At present, natural mixing of fresh and salt water occurs in the area and the effect of cutting off some of the fresh water input, to the outlying Wash, has to be examined. The storms of the North Sea present a threat to the proposed embankments, so their frequency, wind directions and intensities must be considered, together with the storm surges – exceptionally high tides – that occur from time to time in the southern North Sea. The investigation of these and various other questions aims at imposing human control on a large and complex natural environment with a single objective in view – the provision of abundant potable water for man.

In contemplating large-scale intervention in natural systems, of the kind advocated for the Wash, our understanding of the systems needs to be comprehensive. Although perhaps less widely appreciated than the balance of living communities, the physical environment, too, comprises dynamic and fluctuating equilibria. Innumerable physical and chemical factors interact in an extremely complicated way. Some of the basic equilibria, such as the salinity of the oceans and the oxygen pressure of the air, have maintained themselves in stable configurations over very long periods of time. At present, however, man's capacity to disturb these balances is greater than it has ever been, and changes that might take thousands or even millions of years to effect by natural causes can be brought about locally or globally in a matter of decades. In this context the role of the earth scientist is to help to safeguard the physical environment of the planet in a period when unprecedented demands are being put upon it for the wellbeing of man.

References and further reading

P. Cloud and A. Gibor, 'The Oxygen Cycle', *Scientific American*, 223 (1970), 3, pp. 110–23.
P. Cloud, *Resources and Man*, Committee on Resources and Man (San Francisco 1969).
D. R. Coates, *Environmental Geomorphology* (Binghamton 1971).
P. T. Flawn, *Environmental Geology* (New York 1970).

J.C.Frye, 'A geologist views the environment', *Illinois State Geological Survey: Environmental Geology Notes*, 42 (1971), pp. 1–9.

D.W.Hood, *Impingement of Man on the Oceans* (New York 1971).

M. Overman, *Water, Solutions to a Problem of Supply and Demand* (London 1968).

W. A. Vogely (director), *Mineral Facts and Problems, US Bureau of Mines Bulletin*, 630, 4th ed. (1970).

Atmospheric science   4

LIFE IN A
CHANGING CLIMATE
L. P. Smith

The environment contains a large volume of air so that, unless such air receives due consideration, the biosphere shrinks into a biohemisphere. The only major addition to the Earth's energy comes from the Sun by way of the atmosphere; all other energy sources merely convert pre-existing and essentially dwindling supplies. The continuing provision of water essential for agriculture, wildlife, industry, power generation and domestic life depends basically on atmospheric processes. As atmospheric conditions affect all biological events and at times dominate them, no process within the environment can be fully explained unless the accompanying meteorological processes are taken into account. Weather and climate must therefore be interpreted in terms of their effect on plants, animals, the soil and human affairs; conversely, it must not be forgotten that human activities on or near the Earth's surface can have their effect on weather and climate.

Consideration will first be given to the question of acquisition of data and then to three representative aspects of atmospheric science in relation to environmental science – namely, the circulation of pollutants, the complex influences of weather on crops, and changes of climate. In each case, the situation is a mixture of reliable but fragmentary circumstantial knowledge, and uncertain or incomplete theory. The relevance of weather to such problems as human

L. P. Smith is a member of the (UK) Meteorological Office. Until recently he was president of the Commission for Agricultural Meteorology of the World Meteorological Organization.

disease, forest fires (and hence smoke pollution), urban 'micro-climates', and a host of other aspects of life and the environment, must herein be neglected, and also the immense advances in global 'monitoring' by satellite and the less conclusive experiments in deliberate weather modification.

The lesson to be learnt by environmental scientists from those experienced in applied meteorology is that every environment is different and inconstant, and that complex systems can be understood only by meticulous data-collection, logical analysis and repeated practical-scale investigation to identify the governing factors in each real situation. Neglect of any part of this combination leads quickly to nonsense.

Meteorological measurements

The science of meteorology is fortunate in that it has, over many years, carried out a system of monitoring whereby it has acquired a mass of regularly recorded data of a fair standard of accuracy and reasonable homogeneity. This data bank contains certain imperfections and there are a few unavoidable omissions, but nevertheless it constitutes a basic supply of working material lacking in many other environmental sciences. Indeed, all sampling and monitoring procedures now used or envisaged for environmental management imitate to some degree the proved routines of the meteorologist.

At a time when there are arguments about changes in global mean temperature of a fraction of a degree, it is important that non-meteorologists realize the limitations of environmental data-collection. While human and instrumental errors of observation can be largely eliminated by careful checking and by better instruments, there remain two other sources of error which are less obvious and more intransigent. One is that the conditions of exposure of the measuring instrument can greatly affect its reading; the other is the difficulty of adequately sampling an environment which is constantly changing in time and place.

To be able to judge, therefore, how well any series of meteorological measurements represents the true circumstances, the conditions under which the measurements are made must be specified. For example, any air-temperature reading is only the temperature of the sensing element in a given exposure. The standard international practice is to use a mercury-in-glass thermometer within a white-painted louvred screen, but even so, the height of the screen, the nature of the ground below the screen and the siting of the screen varies from country to country. Non-standard measurements are also made with many different thermometers and many different conditions of exposure, and are thus very difficult to compare.

Soil temperatures are less widely taken and, although there is no problem about screening the thermometer from direct sunlight, defining the nature of the soil and its surface condition is not simple and taking the measurement disturbs the soil and hence tends to affect the temperature. Nor can rainfall measurements be regarded as exact, for the sampling errors can be significant and several types of gauge and methods of exposure are employed. The final errors are usually numerically small but they cannot always be disregarded. Errors of snowfall measurement may be substantial. The counterpart of rainfall, namely evaporation, is so difficult to measure accurately that an estimate made indirectly (and intelligently) from other meteorological measurements can often be more reliable.

Recent years have seen a great increase in the accuracy and extent of measurements of solar radiation. Net radiation absorbed, which would be a very valuable parameter for biological purposes, is still difficult to record. Better information exists regarding sunshine hours or cloud amounts and from these data it is often possible to estimate radiation details to a fair degree of approximation. No really efficient recording instrument is yet available for either air humidity or soil moisture and in both cases adequate sampling is hard to achieve. Sampling errors also occur in wind measurements, where there are special difficulties of allowing for rapid variations in space and time.

For all these reasons, care must be taken, when using 'standard' meteorological data in studying practical environmental problems, that the measurements fall within a known radius of uncertainty. As meteorologists still have these difficulties after more than a century of daily practice, it is not inappropriate to ask how reliable and how 'standard' are the measurements being newly taken of pollution and other environmental factors.

World-wide pollution

Pollution emitted from industrial, agricultural, military and other sources, including natural sources such as volcanoes, is transmitted by the wind as well as by water. Airborne pollution, besides being potentially serious on a local scale, creates international problems when it crosses national frontiers or even, entering the general atmospheric circulation, becomes a modifying factor to the climate of the whole planet.

The pollution itself, especially the radioactive materials released from tests of nuclear weapons, has provided atmospheric scientists with ready-made tracers for investigating the circuits of air masses, and for discovering the world-wide fate of the pollutants. Analyses of spore-trap catches are providing a new source of valuable data. Long-life meteorological balloons are also being used to track the movements of air around the world. In addition, the ice caps of Greenland and Antarctica preserve the remains of snowfalls of many centuries and enable chemists to find out how pollution of the atmosphere has changed through the ages.

Several processes determine where pollutants released at one place will finally arrive. One such process is mixing, wherein the turbulence of the air and random diffusion of particles and gases combine to disperse and dilute the concentration of the material. Horizontal winds carry it away from its point of origin and normally, in middle latitudes, the prevailing winds in the lower atmosphere transfer material eastwards, taking about twelve days to circle the Earth; nearer the equator the circulation is more in the westerly direction. Vertical movements tend to carry the material upwards or downwards to altitudes where northerly or southerly winds transfer it to other latitudes. This varying and complex pattern of circulation is, however, slow to transfer material across the equator so that, to a first approximation, pollutants released in either hemisphere tend to remain there.

The existence of many particles and gases in the lower atmosphere is generally terminated by rainfall or snowfall which washes them out of the air and back to the ground; typically this occurs within a week of their release, or up to a month for small particles, thus still allowing ample time for them to reach distant places. Gases and the finest particles which escape this washing-out process can eventually find their way into the stratosphere – the calm, rarefied region above about 10 kilometres altitude. Once they have reached the stratosphere, they tend to remain there for some years. For example, some of the radioactivity injected into the stratosphere during the intensive testing of nuclear weapons in the 1950s is still making its slow way back to the Earth's surface.

The knowledge of physical processes which lead to the addition of pollutants of all types to the atmosphere or to water or soil, and which furthermore control their transfer, dispersal and finally their deposition on new sites, is an essential part of the young science of aerobiology. Pests and diseases are transmissible via the air, like the physical and chemical pollutants, and their survival and spread are largely determined by meteorological factors.

In short, pollutants carry no passports and the atmosphere is an efficient means of transporting them around the Earth. Gases such as the oxides of carbon and sulphur enter into complex physical and biological cycles. To keep the problems thus raised in perspective, however, two points should be borne in mind. In the first place, the very process of scattering the pollutants quickly reduces their concentration at any one point to very low levels. Secondly, by the time one comes to consider the global effects of pollution, as opposed to local and regional effects immediately downwind from the sources, the man-made material is usually small in comparison with the natural material. One volcanic explosion can put more particles into the atmosphere than all the industrial activity of human history. That is not to say we can ignore the global effects of man-made pollution, but much discussion about possible climatic and geochemical consequences is vitiated by our present ignorance of the natural baselines of atmospheric impurities and the range of variations which occur naturally. Regular air-sampling both near the ground and in the higher atmosphere would be worth more than a library of learned speculation on the subject.

Agricultural meteorology

Micrometeorologists deal with the transfer, within the biosphere, of heat, momentum, water, chemically active gases and particles, and with such processes as radiation, conduction, advection, convection, diffusion and turbulence. The full explanation of the relevance of these processes to plant science, animal science and soil science is too intricate to review in a short space, but the whole subject has recently been summarized by a panel of micrometeorologists for the World Meteorological Organization (Smith 1972). Events within the human environment are the results of interacting factors involving biological material, the soil, the weather, and mankind in general. The concentration of the following paragraphs on the physical effects does not in any way imply that the other factors can ever be ignored, least of all those involving human decisions and actions.

A suitable starting point is the germination of the seed, although a fertility cycle has no beginning but only a rebirth; John Barleycorn is never killed and the goddess of fertility herself is (as yet) immortal. If the seed is viable, the emergence of the plant is largely a response to soil temperature and soil moisture, which depend on past and current weather. Survival throughout the crop cycle depends on the current weather which, in turn, is dependent on the general circulation of the atmosphere around the world, and the modifications due to the location. The chief requirement is an adequate growing season, which can be limited by the onset of adverse conditions of radiation, temperature, wind or moisture. The essential part of a crop cycle, pollination, can be affected directly by weather, or else indirectly through the reaction of pollinating insects to the weather.

The effect of the different aspects of weather on the quantity and quality of crops is such a complex question that it cannot yet be answered completely. In certain areas with certain crops, where a dominating meteorological factor is identifiable, fairly reliable semi-empirical relationships between yields and weather can be adduced, yet these are seldom applicable, without further investigation, in places other than their area of origin. These semi-empirical relationships are excellent examples of simplifying a complex situation without sacrificing accuracy, but their want of generality illustrates the besetting difficulty that no two environments are ever quite alike.

The reproduction of animals is also weather-sensitive; conception, pregnancy and birth are all influenced by direct environmental effects, and indirectly through fodder quantity and quality. Direct and indirect factors also affect animal growth and yield of products such as wool and milk. The mobility of animals, which allows some adjustment to the stresses of weather, makes semi-empirical links with meteorological factors more difficult to establish than for plants.

Chance intervenes in agriculture, in adverse weather conditions, whether through damaging storms of wind, rain or hail, or through weather-related catastrophes such as fire and disease. Frost is a common restrictive factor in temperate-zone agriculture, in the early or very late stages of crop growth. It can limit the growing period of a crop or, on occasions, can cause severe losses in a normally safe situation, and the meteorological interpretation of the frequency, intensity and location of frost is correspondingly important.

The effects of weather on the timing and intensity of attack on plants and animals by pests or diseases can be of three broad kinds. Sometimes the pest or disease is critically dependent on particular weather conditions; secondly, the weather may play a more indirect role, for example in creating stresses which influence the liability to attack; thirdly, an epidemic spread may be influenced by wind and other weather conditions, when the pests or pathogens are airborne. Better knowledge of the links between weather and pests and diseases is of great economic and ecological importance, because precautionary or remedial action, by spraying or dusting with biocides, should be restricted to times when it is really needed, so as to avoid unnecessary use of potential pollutants.

Soil erosion is a natural process, in part meteorological in origin, but it can be disastrously accelerated by man's misuse of his soil, with effects that may be virtually without remedy. Erosion and loss of topsoil are brought about by wind or rain or a combination of both, and they occur chiefly in areas where the land use is out of harmony with the climate. Man's stupidity or cupidity has caused erosion on a grand scale in many areas over many centuries. It can be prevented on an equally grand scale by sensible interpretation of the physical conditions of soil and weather. The 'half-life' of many pollutants is measured in days or weeks, but the 'half-life' of erosion is measured in centuries.

Not all of the influences of the physical environment on agriculture are damaging or deleterious, and not all of man's actions are harmful. The two chief improvements which are of a meteorological nature are irrigation and the provision of shelter from wind or sun. Modern irrigation is based, both in planning and in practice, on meteorological assessments and it effectively optimizes the soil-moisture conditions. Shelter, either in the form of living hedges or trees, or provided by the use of materials such as fences or netting, creates better growing conditions over large areas.

Over smaller areas, the use of mulches to cover the soil can bring about many beneficial results, including the alteration of the plant microclimate. In food production, the most specific alterations are made by the use of a cover of glass or transparent plastic; the equivalent change for animals being the provision of housing or shelter. It is ironical that far more care is taken to try to identify and achieve optimum conditions for plants and animals than for human beings.

Any alteration of the Earth's surface, from the hoeing of the soil on a garden plot to the creation of large man-made lakes by the erection of dams, alters the environment in which man lives and in which his food is grown. The consequences can be assessed only by correct understanding of the physical processes involved, and there is no time or space left for the old method of learning by trial and error.

Changes of climate

The most important contribution that climatology has to try to make to environmental planning in the broadest sense is to predict the climate of the years and decades ahead. It should be said straight away that this cannot yet be done, although attempts are continually made. There are grave difficulties even in detecting an actual climatic change, when the very idea of climate is a summation of fluctuating weather, but the climate undoubtedly does vary significantly over periods of several decades, with far-reaching consequences in many countries for agriculture, fisheries, water supplies, transport, hydro-electric power, tourism, pollution control and many other activities.

The global pattern of weather is very complex and a particular climatic trend will improve conditions in some countries and worsen them in others. The most sensitive areas are the marginal ones, in semi-arid or semi-frozen regions, where a slight change in average conditions can, for example, produce six hungry years in ten instead of three in ten. When man is investing in the future more intensively than ever before, to feed, house and employ his growing numbers, it is a matter for concern that we cannot yet tell which of his development plans may be confounded by climatic changes, or which may indeed be helped.

In discussions of climatic change in the context of environmental science, most publicity is given to possible man-made changes of climate, the arguments revolving around the warming effect of additional carbon dioxide produced by burning fuels and the cooling effect of particles added to the atmosphere by the same process and by other industrial activities. These are questions that deserve, and are receiving, careful attention from meteorologists, and they complicate any consideration of present climatic changes. But no human activity was conceivably responsible for the ice ages, nor for the lesser but significant fluctuations of climate that have occurred during the ten thousand years since the last retreat of the ice.

The warm period around the eleventh century AD, for example, which allowed the Vikings to prosper and expand, contrasts sharply with the 'little ice age' of around the seventeenth century when northern Europe suffered a long succession of severe winters. Between 1880 and 1940 the global mean temperature rose by 0.6 degrees C, but before anyone cries 'pollution' it should be added that since 1940 the climate has shown signs of becoming cooler, especially in north-west Europe. There is no consistent global trend and the natural range of fluctuations is great enough to mask the effects of present human activity (Sawyer 1972).

Some examples of effects of climatic change in recent decades can be given for Britain. This is not an obvious marginal region for agriculture as a whole but, as its farmers pride themselves on matching crops very closely to climate, unexpected changes can make specialized crops 'marginal'. Since 1950 extensive rain at harvest time has become more frequent than it was in the 1920s and 1930s, favouring grass (and hence milk and meat) over grain, in keeping with the old saying, 'A year good for grass is good for nothing else'. Without combine harvesters and grain-drying equipment cereal growing would have been badly affected. Another saying is: 'A peck of dust in March is worth a king's ransom' – meaning that dry ground and an early start to cultivation and sowing are necessary for maximum yields. Around 1940 the farmer in south-east England had his 'March dust' five years out of ten; since then it has become a less frequent event, and springs, especially in the hills, have become later, thus shortening the growing season in years which are economically critical. Occurrences of disease of plants and livestock have also followed the relatively small changes of the shifting climate.

With such marked effects in a climatically well-favoured country like Britain, it is not difficult to understand why there is anxiety about the present ice threat to the cod fisheries of Greenland and Iceland, and about the droughts that have recently become commoner in some semi-arid parts of Africa and the Near East.

We cannot tell whether the current climatic trends will continue or if they will reverse. The present hope for climatic forecasting lies in the making of elaborate mathematical models of the Earth's atmosphere and its interactions with the oceans and land masses. When these models become sufficiently realistic we may at last understand why climatic changes occur, what natural changes are in prospect and precisely how and to what degree human activity is influencing the atmosphere.

Knowledge and ignorance

Many are the rewards which await the man or woman who elects to serve at the altar of the goddess of science but, in the harsh world of reality, knowledge is sterile until it is married to practical problems. Problems demand decisions and, although decisions have to be made with imperfect knowledge, they should never be made with no knowledge at all; there are limits to the virtues of intuition.

Some environmental decisions concern short-term action within a roughly pre-determined plan. In biometeorology, they range from questions of whether or not to wear a coat or carry an umbrella, through the choice of time to sow or harvest a crop, to national assessments of a requirement for heating, lighting or air-conditioning. In every case the physical conditions need thought, and the past, present and future weather have to be taken into account. In long-term decisions, for forward planning, climate prediction cannot be evaded. It concerns the future use of our limited natural resources and its issues of environmental strategy and capital investment are at least two orders of magnitude greater than the short-term tactical issues.

Although considerable progress has been made in recent decades it is idle to presume that more than the surface of an extensive field of research has yet been scratched. Investigation in depth at a myriad of points of application is urgently required; so are efforts to bring the complexities of meteorology into correct juxtaposition with the complexities of other disciplines. It is rare that a single brain can sufficiently comprehend the necessary matrix of thought. Team work, mutual collaboration and consultation are therefore essential, and the habit of co-operation must start in the early stages of education. Perhaps the hardest things to learn are the limits to one's own knowledge, and when and how to obtain outside assistance. It is a wise man who realizes how little he knows; an even wiser man knows to whom to turn to remedy the gaps in his own repertoire. In the environmental sciences, of which meteorology is one of the oldest, the days of the soloist are over; the age of the orchestra is slowly dawning.

References and further reading

A. G. Fosdyke, *Meteorological Factors in Air Pollution*, WMO Technical Note 114 (Geneva 1970).

Reginal E. Newell, 'The global circulation of atmospheric pollution', *Scientific American*, 224 (1971), 1, pp. 32–42.

J. S. Sawyer, 'Man-made carbon dioxide and the greenhouse effect', *Nature*, 239 (1972), pp. 23–6.

L. P. Smith, *Weather and Animal Diseases*, WMO Technical Note 113 (Geneva 1970).

L. P. Smith (ed.), *The Application of Micrometeorology to Agricultural Problems*, WMO Technical Note 119 (Geneva 1972).

Unesco, *Changes of Climate*, Proceedings of the Rome Symposium (Paris 1963).

FARMING AS ENVIRONMENTAL TECHNOLOGY

A. H. Bunting

Agriculture first appears in the archaeological record about ten thousand years, perhaps four hundred generations, ago. This was the great divide in the history of our relations with the natural environment. Agriculture enabled our ancestors increasingly to manage and control the world about them, and so to escape from almost total domination by the environment. It led at once to widespread and far-reaching environmental changes. Men had of course lived through vast changes before: ice sheets came and went; forests and deserts waxed and waned. But since farming began, virtually all the primary vegetation of the Earth has been modified by man so that today most of it, including much of the tropical rain forest, is secondary growth. The treasured landscapes of Britain, like the woodlands and forests of much of the north-eastern United States, are artefacts of past agricultural change. The often-invoked 'balance of nature' is man-made, although the changes have included not a few ecological disasters. Irrigation has sometimes extended salt deserts; the scars of the 'Dust Bowls' of the 1930s can

A. H. Bunting is an agricultural botanist and professor of overseas agricultural development at Reading University. He is a member of the United Nations Advisory Committee on the Application of Science and Technology to Development (ACAST) and is associated with the International Institute of Tropical Agriculture, Ibadan, Nigeria. He was for some years joint editor of the *Journal of Applied Ecology*.

still be seen in parts of the United States; and the East Anglian Breckland owes its unique ecological character to irreversible deterioration of the soil induced by ancient agriculture.

Agricultural ecosystems are usually far more productive than wild systems, particularly when environmental constraints are lessened by technology. Sugar cane in Hawaii, irrigated and adequately fertilized in a two-year cycle, can accumulate more than seventy tons of dry matter per hectare per year, of which four-fifths may be sugar. Most wild vegetation falls short of one-tenth of this productivity. But agricultural systems, like all secondary successions, are unstable. The farmer and the agricultural ecologist work in a dynamic ecology of change, rather than one of stability.

Adaptation: a case-study of farming ecology in northern Nigeria

To modify the major features of the environment is either impossible or costly. It is usually simpler and cheaper to fit species, varieties and breeds of crop plants and farm animals, and farming systems as a whole, to the principal features of the 'natural' environment, as most traditional farming systems do.

In the northern states of Nigeria, the climate becomes steadily wetter from north to south, but there is a pronounced dry season everywhere. Towards the end of the dry season a 'hungry gap' begins in most years, as the previous season's produce starts to run out. At the beginning of the rains much of the soil nitrogen becomes soluble and it may be rapidly leached from the upper layers of the soil, or else wasted on weeds if sowing and weeding are delayed.

To these dominant features of their environment, the farmers have matched their crops and farm systems with remarkable precision. Timing is all-important, and it depends on judicious use of both 'short-day' crops, which are photosensitive and respond to slight changes in the duration of daylight, so that they flower on a more or less precise date, and 'day-neutral' crops, which are not so governed.

In the drier tracts, down to about 9° N latitude, the staple crops are cereals – bulrush millets (*Pennisetum typhoideum* and other species) and particularly sorghum (*Sorghum bicolor*). Like most other cereals, these contain enough protein, of sufficiently satisfactory biological value, to satisfy the protein needs of populations which eat enough of them to meet their needs for energy. As early

49

as possible in the rains, which may start at any time from mid-April to mid-June, farmers in the wetter parts of the cereal zone sow bulrush millet at wide spacings. The day-neutral variety they use flowers (like a desert ephemeral) at a fixed and small leaf-number, and so gives the earliest possible harvest to shorten the hungry gap.

The main cereal crop, sorghum, is sown rather later, between the rows of millet. In the less arid areas the local sorts of sorghum are very delicately short-day photosensitive, and they flower at times strictly related to the average dates of the end of the rains, which are fairly sharply defined, in their own localities. The precise relation is important: insects and moulds will damage the ears and grains of a form which flowers too early, while one which flowers too late will exhaust the residual water in the soil before the grains are filled.

After the sorghum has been sown it usually has to be weeded twice: this creates the peak labour demand of the whole year and limits the total amount of land a family can handle, and so the amount of productive work the farm system can offer at other times of year. After weeding, many farmers undersow a legume crop (the cowpea *Vigna unguiculata*) among the sorghum. The varieties they use are spreading, can tolerate shading by the sorghum and are photosensitive in a complex way; inflorescences can be *initiated* at any daylength which occurs in northern Nigeria, but will *expand* to bring the crop into flower only towards the end of the rains, as the leaf area of the sorghum declines. In this way, the timetables of the two crops are physiologically coupled. The cowpea crop covers the soil surface, protecting it from beating rain, and controls weeds. Being a legume, it may contribute some nitrogen to the system, at a time when there is little to be had otherwise, and it yields a small crop of protein food into the bargain.

Further north, short-day photosensitive bulrush millets, adapted to the drier conditions, replace the short-day sorghum; and further north still, where both the start and the duration of the rains are very uncertain, day-neutral, early sorts of both sorghum and bulrush millets are the main crops.

In the wetter south, below 9° N, the staples are the yams (*Dioscorea* spp.), cassava (*Manihot utilissima*), cocoyam (*Xanthosoma*) and taro (*Colcasia*). The starchy foods from these crops contain virtually no protein, but one or other of them can usually be harvested in most months of the year. Consequently, in the hungry gap, people are short of protein rather than energy, and, sure enough, we find that the early crop sown to break the gap is not a cereal, but an upright, day-neutral, early-flowering form of cow-pea, often sown by itself, which can start to yield protein-rich food as little as eight weeks after sowing.

All these precise physiological relationships have evidently been selected, albeit unconsciously, to fit the timetables of the crops to their economic and social roles in the varying and largely un-predictable conditions of the different climatic zones.

Nigerian farmers also grow cash crops, which compete with the food crops for land, labour and time. Many farm families have to divide precious early-season labour, in the short period before leaching decreases the amount of nitrogen available in the surface layers of the soil, between cotton and the staple food crops on which they depend for survival. Not surprisingly, cotton comes out the loser and is planted too late to yield more than perhaps a fifth of its potential.

Only recently have we begun to breed the new varieties, and to devise the new agronomic and protection systems, that will enable farmers to obtain better returns from their efforts on later-planted cotton. Another way of lessening the competition for labour may rest on some recently bred 'high-yielding' varieties of sorghum, with which farmers might be able to decrease to a half, or even a third, the area that they devote to this crop, so releasing land, time and labour for early sown cotton. The second major cash crop, the groundnut, is a legume and so escapes the seasonal constraints of nitrogen supply. It can consequently be sown later, after the staple foods have been weeded.

Farm ecosystems: inputs and processes

The principal natural inputs into all ecosystems are water, carbon dioxide and solar energy – which, through the process of photo-synthesis, provide the materials and the biological energy sources which enable green plants to grow – and the essential chemical elements we think of as plant nutrients. The rates of growth, and the yields and other outputs, depend on the processes in which these inputs are used.

CARBON DIOXIDE AND WATER. Carbon dioxide enters plants through their stomata and, as long as these small apertures are open, water vapour diffuses through them into the atmosphere.

Consequently, assimilation of carbon and loss of water by transpiration are intimately linked. Transpiration is a largely passive physical process – a special form of evaporation – affected by the net income of solar energy and the physical characteristics of soil, plant and atmosphere. The amount of water lost from vegetation which fully covers the ground and is freely supplied with water depends mainly on land area and time, and is relatively little altered by the number or nature of the individual plants.

When water is less freely available, so that it does not enter the roots fast enough to supply the needs of transpiration, leaves become less turgid and the stomata tend to close, and so to lessen the rates of both transpiration and assimilation of carbon dioxide. Transpiration helps to cool the leaves, which may be permanently damaged if the weather remains clear, dry and hot for very long. In dry situations, the best chances of ecological or economic success therefore lie with short-lived plants or crops which can complete their lives before the water available to them is exhausted, particularly if they have deep roots to increase the supply by tapping whatever stores of water there may be lower down in the soil.

For the relatively simple case of a single-species crop, all these relations can be realistically modelled as a network of potentials, resistances and consequent flows of water and water vapour from soil to atmosphere. For over twenty years such models have been used in a simple form, using weather data, to estimate water needs in irrigation and to account for the observed relations between land use, vegetation and hydrology. From the same basic ideas we have also derived a clear and quantitative understanding of the essential differences, for crops and other vegetation, between temperate and tropical environments. For particular crop species in particular places it is possible to relate growth quantitatively to the seasonal course of weather and climate. Such models have accounted for over 90 per cent of the variation in the growth rates of grass in England, and of peanuts in Nigeria.

ENERGY. Most of the radiant energy received by vegetation is used to evaporate water, and heat the air and soil. Ten per cent or so of it can theoretically be used in photosynthesis, and up to 4 per cent has been recovered as dry matter over short periods under experimental conditions. In practice, however, no more than an average of about 0.3 per cent is recovered as chemical energy even in advanced agriculture; in much of the traditional agriculture of the tropics, the average is o.o6 per cent only.

There are many reasons for this seeming inefficiency. One is inadequate leaf area, in crops which are widely spaced or take a long time to cover the ground. Another is the phenomenon of light saturation. In most species, leaves held at right angles to the incoming radiation reach a maximum rate of assimilation at intensities well below those of full daylight. Many advanced crops and crop varieties dispose their uppermost leaves at appropriately slanting angles to the noon light while their more shaded lower (and so, usually also older and less efficient) leaves are held more nearly horizontally. We can now design crops and crop mixtures so that their 'architecture' maximizes total productivity and yield, as in the new wheat and rice varieties of the Green Revolution.

Maize, sugar cane, sorghum and some other tropical species (many but not all of which are grasses) appear to absorb carbon dioxide more rapidly than other species, in warm, sunny conditions. This is largely because in these species, unlike most temperate ones, light does not increase the rate of respiration in their tissues. Such plants may offer a road to larger yields but, if so, it is only one of many. In most crops, yield is limited less by the rate of increase of dry weight per unit of green area than by the size and duration of the green area, by the way the timetable of the crop is adapted to the length and nature of the growing season, and by those inherited physiological and morphological attributes that determine the potential size of the economically useful parts.

NUTRIENTS. Most of the nitrogen in the world's soils, waters, plants and animals is derived ultimately from the biological fixation of the nitrogen of the air by micro-organisms which are either free-living or symbiotic with plants such as legumes. They can fix up to several hundred kilograms of nitrogen per hectare per year. All the energy they use for this purpose is ultimately derived from the sun by way of green plants. The management and improvement of these nitrogen-fixing microbial systems is a most important sector of agricultural science.

In many circumstances, the natural supply of nitrogen is not large enough for maximum growth and yield. Consequently, alternative sources of readily available nitrogen are added to part at least of the soil area in all but the simplest farm systems – more or less decomposed organic wastes, animal bedding, household refuse, naturally occurring nitrates, and, during the past half century, manufactured nitrogen compounds, the so-called fertilizers. (Many of the older methods are no more than a redistribution, which concentrates, on a small area, nutrients drawn from a larger one.) The available supplies of phosphate and potassium may also limit the productivity of crops and here, too, supplements are generally used.

Shortages of minor nutrients, for example sulphur, iron, molybdenum, copper or zinc, may restrict growth and yield on particular soils, even though they are needed only in small amounts. Particularly striking effects have been produced (classically in the Ninety Mile Desert of South Australia) by additions as small as a few hundred grams to the hectare of appropriate compounds of these 'trace' elements.

These nutritional improvements, applied to new varieties that can take full advantage of them, are among the most important reasons why modern agricultural vegetation is so much more productive than wild vegetation in the same place.

BIOLOGICAL RECYCLING. As vegetation grows, whole organisms die, parts of organisms are shed, and waste or surplus materials are excreted. They become the raw material for a series of degradative changes whereby micro-organisms obtain chemical energy, including that used to fix nitrogen. In these processes nitrogen, phosphate and other primary chemical inputs once more become available for use by green plants, and carbon dioxide returns to the atmosphere.

The micro-organisms which make their living in the degradative phase are so versatile and resilient that they can often break down unusual and artificial molecules, including those of many farm chemicals, plastics and other unaccustomed materials.

One of the products of microbial action in the soil is a gas, ethylene, which actively regulates the growth of roots and other parts of plants. In conventional soil analysis the samples are air-dried and sieved, so that any ethylene it contains is lost. Who knows what other volatile substances, active in trace amounts, may so far have escaped the sceptical soil chemist's net?

Farm ecosystems: competition, distribution, output and change

In more or less stable vegetation the output of the degradative phase is equivalent to the inputs into the growth phase; net productivity is zero. Economic productivity may nevertheless be considerable. Thus in perennial plantations, annual input is divided between economic material, new growth and degradation. In annual crops, in the annual shoots of some perennials, and in many secondary successions, seasonal accumulation may greatly exceed degradation. Moreover a large fraction of the excess may be economically useful. The output of food, fibre and other useful products from advanced crop varieties often exceeds half of the season's net accumulation of biomass.

The distribution of the total increment in biomass between different green plants in a community is determined by competitive or interference processes, particularly 'competition' for light. In wild communities fine details of timing, particularly of leaf expansion, enable mixtures of species to coexist, and in traditional farming suitable mixtures may yield more than equivalent areas of pure cultures. Many crop plants have horizontally spreading leaves which enable them to compete with weeds for primary biomass, but advanced, large-yielding crop varieties which grow faster because they have inclined leaves often control weeds much less effectively. The ways in which economic plants accumulate dry matter in special organs – fruits, ears, corms, roots, tubers, trunks – also tend to make them less fit to survive in competition with wild plants.

Protected by man from competition, however, many crops have enormously increased their ranges of environmental adaptation. Potatoes from the high Andes and wheat from arid south-west Asia are grown today from the equator to the Arctic circle, while maize from semi-arid Mexico, having traversed the whole of the tropical world, is pressing hard on their heels in northern Europe.

Competition between different species leads to succession, towards increasingly stable communities in which taller and perennial plants tend to replace shorter annuals. In the process, the inputs accumulate in the larger plants and the weedy pioneer species are eliminated. When the vegetation is buried or burnt the accumulated nutrients, and the absence of many diseases, pests and weeds, favour the growth of crops. This is the rationale not only of the traditional slash-and-burn systems of farming in many of the less-densely populated parts of the developing world, but also of the alternate husbandry or ley rotations of temperate farming systems.

The secondary phase of distribution, through grazing animals, can also be very important in cycles of this sort. The herbivores take

55

in carbon and nutrients, 'burn' off much of the carbon as carbon dioxide, and recycle those nutrients (including much of the nitrogen and phosphate) which they do not need for their growth. Essentially similar processes go on in the food chains of wild communities, and indeed the concepts and methodology for studies of energy flow and conversion losses in wild systems are largely derived from agricultural science.

A very wide range of pest and disease organisms – fungi, bacteria, viruses, insects, arachnids, nematodes, mammals and birds – compete with man for energy and materials in agricultural systems. Much agricultural science deals with the biology of this class of secondary relationship, with the breeding of plants which resist or are immune to attack, and with devising systems of farming in which losses are minimized. In perennial crops in ecologically isolated situations (glasshouses, islands, oases) the production systems can be made so stable and predictable that pests and diseases are held in check entirely by biological means, but most farming is not, and in a continually changing world probably cannot be, organized in this way. Direct methods, including the use of specific biological or chemical control materials, are therefore likely to be necessary components of pest management systems for many years to come, though the materials, and the ways in which they are used, will surely continue to improve, as they have done so spectacularly in the past in human and veterinary medicine.

Man: the ecology of rural development

One of the main conceptual weaknesses of much professional ecology in advanced countries is that it tends to neglect, exclude or even reject the human species, except perhaps as a reprehensible source of interference, damage and disaster. In agricultural ecology, especially in the developing world, human beings in large numbers are at the centre of the biological action, in intimate interplay with wild and cultivated species. Agricultural ecology is consequently an integral part of the social ecology of man.

The principal social and political task of our evolving species in the years ahead is to lessen poverty and inequality, both within and between nations. In a rich and industrial nation, it may be possible to hope that this will be largely a question of management and distribution. The task is far more complex in a poor country where the great majority of a rapidly growing population depends solely on agriculture for survival and is likely to continue to do so for many years to come, urban growth notwithstanding.

Since most existing systems seem to use environmental and human resources with considerable efficiency – as the Nigerian example may suggest – new inputs, and new technological processes, are usually necessary to improve them. Simple rearrangements are seldom enough to make systems more productive. Nor are 'single factor' solutions normally effective: a package of new inputs and processes has usually to be devised, tailored to the objectives, resources and opportunities of farm communities and families.

Moreover, to increase productivity is not enough either: it has to be done in labour-intensive ways which also increase the total volume of productive work available to the rural society, particularly if it is increasing, as many are, in absolute numbers of people. Further, the rural change process has to be consciously and effectively articulated with planned change in other sectors of the economy, and particularly with the growth of non-agricultural economic activities and of economic exchanges between the agricultural and non-agricultural sectors. Much of the non-agricultural activity in developing countries uses agricultural raw materials; and it seeks in the majority rural sector its main internal markets, both for purchased agricultural inputs and for consumer goods.

Most of the poorer countries are in the tropics. Both their environments and their social systems are significantly different from those of the temperate countries. The advanced farm technology of the developed world is seldom appropriate to these very different circumstances. Technically, it is capital-intensive and labour-saving, it uses purchased inputs on a considerable scale and it has to serve very different social and economic objectives. Hence new, locale-specific methods of farming are needed, not only to use the tropical environment more effectively, but also to meet the particular social and economic needs of particular countries, regions and districts. They must be adapted simultaneously to their human as well as their natural environments. This is the great challenge to agricultural ecology in the developing world.

References and further reading

A. H. Bunting, *Change in Agriculture* (London and New York 1970).

A. H. Bunting, 'The ecology of agriculture in the world today and to-morrow', *Pest Control Strategies for the Future*, National Academy of Sciences, Washington, DC (1972).

D.L.Curtis, 'The relation between the date of heading of Nigerian sorghums and the duration of the growing season', *Journal of Applied Ecology*, 5 (1968), pp. 215–26.

J.D.Eastin, F.A.Haskins, C.Y.Sullivan and C.H.M.van Bavel (eds.) *Physiological Aspects of Crop Yield*, American Society of Agronomy and Crop Science Society of America, Madison, Wisconsin (1969).

M. D. Hatch, C. B. Osmond and R. O. Slatyer, *Photosynthesis and Photorespiration* (London and New York 1971).

H.L.Penman, *Vegetation and Hydrology*, Commonwealth Agricultural Bureaux, Farnham Royal (1963).

H.L.Penman, 'The earth's potential', *Science Journal*, 4 (1968), pp. 43–7.

E. W. Russell, *Soil Conditions and Plant Growth*, 9th ed. (London and New York 1961).

R. O. Slatyer, *Plant-Water Relationships* (London and New York 1967).

W.M.Steele, 'Cowpeas in Nigeria', Ph.D. thesis, University of Reading (1972).

D. F. Westlake, 'Comparisons of plant productivity', *Biological Review*, 38, pp. 385–425.

Anthropology 6

THE HUNTERS' ECOLOGY
Asen Balikci

People born and raised in the contemporary urban environment do not find it easy to grasp that for 99 per cent of his past man has been a hunter. The hunting way of life produced the basic biological and psychological characteristics that made man a success and that we retain. It was as a hunter that man first occupied practically every corner of the continents, yet we know all too little about how he adapted to his environment. The archaeological evidence is limited by its very nature. However, stone tools and bones can tell us what kind of species the hunters preferred and in a few sites we can infer how and where they hunted.

The only complete occupational unit excavated from the Middle Pleistocene period is at Olorgesailie in Kenya. It gives no direct evidence regarding the hunting methods employed by Acheulean man but the rich bone remnants of baboons attest to communal hunting, and this without bows (Isaac 1968). The baboons in a troop were probably surrounded at night, dislodged and clubbed to death by methods similar to those employed by recent Hadza hunters of northern Tanzania.

During the Upper Paleolithic and especially the Mesolithic periods a substantial enrichment and diversification of the hunting cultures took place. The invention of the boat allowed use of marine

Asen Balikci is an anthropologist at the Université de Montréal. He has studied the Netsilik Eskimos in great detail and has recorded their disappearing culture on film.

and river resources; it made travel easier, as did the sledge. The domestication of the dog together with the invention of the bow and arrow must have greatly increased the efficiency of hunting. Grinding stones give evidence of gathering activities. These advances widened the subsistence base and made settlement in villages possible, together with higher population densities. But we are at a loss to know the exact territorial and social organization of ancient hunters, their population policies or their attitudes to their environmental problems.

Dotted around the world today there are surviving groups of hunter-gatherers. Obviously they are a much richer source of evidence about the hunter's ecological adaptations. Yet, by the time trained anthropologists began field research among them, there were very few hunting societies left, usually in uncomfortable environments in isolated refuges (Murdock, 1968, lists twenty-seven such areas). Moreover, the culture and especially the technology of most of these hunting groups were already modified, directly or indirectly, by the influence of neighbouring tillers or herders or European agents. To make matters worse, early field investigators were not interested in studying the strategies of adaptation to the environment or in quantifying their observations; at the time, they lacked a theory of cultural ecology. Despite these difficulties, we can nowadays draw a general picture of hunting as a successful ecological adaptation.

Most remarkable is the extreme variety of natural environments inhabited by recent hunter-gatherers. The Eskimos occupy the vast cold deserts of northern North America and coastal Greenland; the Shoshone Indians live in the arid zone of the Great Basin; the North-West Coast Indians have settlements along the bays and rivers of British Columbia and adjoining areas covered by dense temperate forests; and the Indians of Tierra del Fuego survived in the rainy and inhospitable tip of South America. In Africa, to give just two instances, the Pygmies of the Congo hunt large game in the tropical forest while the Bushmen of the Kalahari Desert are essentially gatherers of wild plants. In Asia the Semang hunt primarily small animals in the tropical jungles in the interior of the Malay Peninsula; the Andaman Islanders exploited the resources both of the sea coast and of the dense jungles. Before the arrival of the Europeans, the Australian aborigines survived by hunting and gathering in the many different climatic zones of their continent.

The comparatively simple lives of these hunting people, and the social systems and practices that they have evolved to cope with their environmental circumstances, provide analogies – caricatures, almost – to our own predicament. A special urgency is given to investigations by the fact that the last of the hunters are vanishing and there is little time left to learn what we can from them.

These hunting societies have plainly developed different adaptive strategies in relation to such widely varying resources. Yet judicious comparisons of certain important aspects of ecological adaptation and social organization may lead us to generalizations applicable to some, if not all, hunting-gathering societies. With this objective in mind let us compare the Netsilik Eskimos and the Kung Bushmen.

Hunters of seal and caribou

The Netsilik Eskimos inhabit a vast area along the Arctic coast of Canada, north-west of Hudson Bay. The country is barren and marshy, covered with rocky hills, tundra flats, lakes, rivers and sea inlets. The climate is rigorous, characterized by long, extremely cold winters and short, cool summers. Half a century ago a Danish explorer, Knud Rasmussen, counted 259 Netsilik Eskimos, of whom 150 were males and 109 females. The 'tribe' was divided into four or five bands occupying different areas. They hunted caribou, musk oxen and seals and were very active fishermen. With the exception of a few berries, the diet consisted entirely of meat and fat. How did the Netsilik Eskimo adapt to this extremely harsh environment? I have to put the question in the past tense because he is abandoning his traditional way of life.

He was a remarkable technologist. With the meagre resources of the environment he was able to elaborate a complex material culture. With snow he built dome-shaped igloos, sleeping counters and kitchen tables. For windows he used ice. From caribou fur skins the Netsilik woman tailored warm winter clothing and with sealskins she made tents, kayak covers, containers and thongs. Since driftwood was very scarce, sledge runners were also made of folded sealskins. Bone, antler, horn, an occasional piece of driftwood and imported iron served for fabricating tools, bows, arrows, harpoons, spears, tent poles and kayak frames. Finally soapstone was all-important, because oil lamps and cooking pots were carved in this material. Thus the specific qualities of a small number of locally available raw materials were ingeniously used for the manufacture of a variety of specialized artifacts, well matched to the prevailing environmental conditions.

With this equipment the Netsilik Eskimos could exploit their food resources flexibly. They spent the winter on the flat ice in large igloo communities which frequently comprised more than sixty individuals in several extended families. These large establishments exploited the ringed seal's habit of keeping open a large number of breathing holes through the thick sea ice; the larger the number of harpooners, each attending to a particular breathing hole, the greater the chance for a speedy catch. Winter was the time of intense social life, visiting, playing, drum-dancing and the great shamanistic performances.

In spring, the pressure of the environment relaxed and the large Netsilik communities often split into two distinct camps. Collective seal-hunting continued at the breathing holes although individuals would sometimes stalk the seal. The excess blubber was stored in stone caches. When the ice broke up in July the kayaks were covered with fresh skins and summer migrations inland began. Each extended family, by itself or in company with another family, travelled deep inland. At stone-weir fishing sites, in early August, fish was speared collectively and new food caches made. In late August, migrating caribou herds were driven by beaters into narrows where they became easy prey for fast-paddling kayakers armed with spears. Besides this co-ordinated group hunt, there was an individualistic alternative – stalking the caribou with bow and arrow. In the fall, the small summer groups would gather again along a river to spear salmon through the thin autumn ice. This was a busy period for the women sewing new winter clothes. Before the spring and summer food caches were exhausted the band reassembled again for the winter seal hunts.

In this dual adaptation, with its winter-marine phase of seal-hunting and its summer-inland phase of caribou-hunting and fishing, there is a clear correlation between the movements of people and the known distribution of game, which made the food supply reasonably predictable. Yet hunting is a highly complex and chancy business and the Netsilik Eskimo was always looking for alternatives, with his practice of substitution and *ad hoc* inventiveness constituting another class of adaptive strategies. For every kind of group hunting there was an individualistic alternative and whenever a particular kind of game failed to appear there was another to fall back on. Frequent fission and fusion of social groups occurred and were clearly adaptive to environmental circumstances.

Sharing of food was extremely important among the Netsilik Eskimos. During childhood a mother would select for her small son about a dozen sharing partners, all non-relatives, for the winter seal. The partnerships endured for life and whenever a hunter killed a seal it was butchered in fourteen parts, each one distinctly named and given to a sharing partner. In summer, when families were living on their own, sharing was much more informal.

Living under conditions of continuous environmental pressure, the Netsilik practised population control. The most effective practice was female infanticide. On King William Island in 1923, Rasmussen counted thirty-eight girls killed out of ninety-six live births in a genealogical sample of eighteen marriages. Since the decision to kill was taken within the family circle, infanticide was evidently adjusted to the providing capacity of the hunter and females were considered to be relatively unproductive. The Netsilik also killed invalids and the elderly.

The nomadism of the hunting life in this extremely rigorous climate imposed a constant strain. Hunting never ceased, the game had to be brought to camp at all costs and the hunter was expected to stay out until a successful kill was made. Despite all their adaptive strategies, starvation occurred among the Western Netsilik, dangerously reducing the size of bands.

Gatherers of the mongongo nut

By contrast, the Kung Bushmen living in the hot and semi-arid Kalahari Desert of South-West Africa seem to enjoy an affluent existence. At least, this is the conclusion reached by Richard Lee (1965, 1968, 1972), who has made detailed surveys of Kung subsistence patterns.

The Bushman's technology is comparatively simple. Huts are made of brushes, women use a garment of tanned skin and three tools suffice for gathering wild plants: a skin container, a sharpened digging stick and nut-cracking stones. The hunting kit is richer; it comprises about twenty weapons and devices, including bows, poisoned arrows and spears. Thus equipped, the Kung succeed in exploiting the relatively rich food resources of their arid environment.

The Kung know of eighty-five species of edible food plants, but the mongongo nut accounts for half of the vegetable diet by weight; this nut is drought resistant, widely distributed and very rich in calories and protein. When Lee asked a Bushman why he hadn't taken to agriculture he replied: 'Why should we plant, when there are so many mongongo nuts in the world ?'

Of the many animal species which are permanent or temporary residents in the area, only ten seem to be regularly hunted by the Bushmen: wart hog, kudu, duiker, steenbok, gemsbok, wildebeeste, spring hare, procupine, antbear and hare. Insects do not play an important role in the diet, although two species of beetle supply poison for arrow points. Bushmen can usually afford to select only the tastiest species and ignore most birds, small mammals, reptiles and insects.

The Kung women are exclusively gatherers of plant food. In the Dobe area, for instance, they go out in groups of three to five and collect for a day or a few hours and always return to camp before evening. The men are primarily hunters although if their hunting fails they may collect some wild plants on their way back to camp. The hunting parties typically consist of just two men, rarely three, and quite often a hunter will go off alone. Dogs add very substantially to hunting efficiency. Unlike the women, the hunters may spend the night in the bush. Unfortunately Lee gives little information on the organization of the hunt. The bow and arrow are used only against large game, with the spear and club to finish off wounded or cornered animals. Wart hogs, antbears and porcupines are asphyxi-ated with smoke in their burrows. Rope snares are set to trap small bucks.

Vegetable foods comprise from 60 to 80 per cent of the total diet by weight; the Kung Bushmen are primarily gatherers and only secondarily hunters, with the women's contribution in the food quest being of primary importance. The constant availability of wild plants means there is no need to store any surplus and the sub-sistence base is so rich that women and men spend no more than two or three days a week in collecting activities.

Lorna Marshall (1960) asserts that the Kung Bushmen of the Nyae Nyae region organize themselves in two groupings and only two – the family and the band, which consists of a number of families. Each band has a headman who co-ordinates the wanderings of the various families. Bands vary in size from fewer than ten individuals to more than fifty. Each band has admitted rights to a water-hole and its surrounding territory, separated from nearby bands by a no-man's-land. For the Dobe area the camp territory extends for six miles around from the water-hole (Lee 1968). Infanticide is also practised, apparently in order to space out the children, and there is no discrimination against female infants.

The Bushmen move their camps five or six times a year, always within a limited area. In the rainy season they may venture ten or twelve miles from the home water-hole; in the dry season they remain near the permanent source of water. In this sense the Kung do not lead a truly 'nomadic' life. On the other hand, they move freely between bands. On average an individual spends only a third of his time living with close relatives, a third visiting other camps, and a third entertaining visitors from other camps (Lee 1968). This pattern of residential flexibility is well supported by the Kung kinship in which any individuals with the same names are deemed to be related. 'Name relatives' receive generous portions during the meat-sharing process, which clearly equalizes meat distribution. More importantly, the system allows for several bands to concentrate for survival around a particular water-hole in drought years (Lee 1972).

Eskimo and Bushmen – some features in common

Let us now compare these two cultures, Netsilik and Kung, and see if we can learn something of a more general nature about the ecological adaptation of hunter-gatherers. A central characteristic of both Eskimo and Bushman societies is co-operation. The first and most obvious pattern of co-operation that they have in common is the division of labour between men and women and the economic inter-dependence of the basic collaborative unit, the husband-wife team. In both societies housekeeping and child-rearing fall into the woman's domain, yet the Bushman women remain remarkably mobile in performing a vital function as provider-gatherers.

More extended forms of collaboration are evident in both societies, in getting, distributing and consuming food. The Eskimos have to hunt together or starve together and, in their case, many collaborative patterns are directly determined by technological factors and by the distribution and movements of the big game. Among the Bushmen, the collaboration in hunting and gathering occurs yet it does not seem to be so imperative. While the Eskimos have rigid rules for sharing seal meat, the Bushmen's food distribution is much more flexible; visitors simply partake in meat feasts. In a sense meat does not circulate among humans, it is the humans that circulate around the meat. Eating together appears to be a corollary to communal meat-sharing.

The crucial point is that in neither society can kinship alone provide the organizational framework for the all-important meat-sharing. Artificial social networks have to be established for connecting non-relatives and equalizing the distribution of game. Anthropologists have traditionally assumed that 'primitive' societies are organized on the basis of kinship. The Netsilik and Kung show that this is not necessarily true and that additional integrative mechanisms have enormous adaptive value for the survival of a hunting society.

Both Netsilik and Kung societies show flexibility in residential groupings. Bands are generally small and of variable size and composition. At almost any time a family can leave for an indefinite period and return later. In both cases, annual cycles of concentration and dispersal reflect the ecological imperatives. In times of stress society does not obey a rigid order but tries to take advantage of changing opportunities. As we have seen, both societies regulate their populations by infanticide, but further comparative research is needed to clarify the role of this drastic practice in the socio-demographic evolution of band societies.

Finally, both Netsilik and Bushmen have an extremely detailed knowledge of their total environment. This includes the distribution of both useful and non-useful flora and fauna according to seasonal changes; the behaviour and anatomy of game species; a vivid sense of topography; climatic phenomena; and so on. Their taxonomies, their specialized vocabularies for the socially important elements of nature and their incredibly numerous place-names provide good evidence of the hunter's world view. It is this store of knowledge that allows the hunter to predict the whereabouts of game with a reasonable degree of certainty. Further, he shares his knowledge with others and there is a continual circulation of new information within the band. The hunting band thus represents a total culture matched to a total environment, and every individual is well acquainted with both.

To the layman the Netsilik and Bushmen are very different

indeed. The first are big-game hunters living in the cold Arctic, the second are gatherers roaming in the bushes of the Kalahari. The differences are real enough, yet they share certain important traits in social organization, in band structure and in the techniques of ecological adaptation. It would be rash, though, to generalize and assume that all hunter-gatherers fit the same pattern. Ecological adaptation is a dynamic and complex process, influencing cultural growth in a variety of ways, and some cultural developments cannot easily be explained by ecological considerations – the extraordinary complexity of Australian forms of kinship is a case in point. Moreover, a particularly rich environment, like that of the North-West Coast Indian hunters and fishermen, may provide a base for very elaborate cultural patterns that bear little resemblance to those of the Netsilik or Bushmen. The very simplicity of these two societies is important for analytical purposes, however, because they allow the observer to abstract and evaluate more directly the creative processes of human minds confronted with the complexity of the natural environment.

References and further reading

A. Balikci, *The Netsilik Eskimo* (New York 1971).
G. L. Isaac, 'Traces of Pleistocene hunters: an East African example', *Man the Hunter*, ed. R. B. Lee and I. DeVore (Chicago 1968).
R. B. Lee, 'Subsistence ecology of Kung Bushmen', Ph.D. thesis, University of California, Berkeley (1965).
R. B. Lee, 'What hunters do for a living', *Man the Hunter*, ed. R. B. Lee and I. DeVore (Chicago 1968).
R. B. Lee, 'Kung spatial organization: an ecological and historical perspective', *Studies of Bushmen Hunter-Gatherers*, ed. I. DeVore and R. B. Lee (in preparation).
R. B. Lee and I. DeVore (ed.), *Man the Hunter* (Chicago 1968).
L. K. Marshall, 'The Kin terminology system of the Kung Bushmen', *Africa*, 27 (1957), p. 1.
L. K. Marshall, 'Kung Bushmen Bands', *Africa*, 30 (1960), p. 325.
G. P. Murdock, 'The current status of the world's hunting and gathering peoples', *Man the Hunter*, ed. R. B. Lee and I. DeVore (Chicago 1968).

THE CULTURAL LANDSCAPE
Grahame Clark

When men all over the world are becoming increasingly aware of threats to their environment, it is perhaps unnecessary to insist that these issue not from basic trends in the natural order but from human activities themselves. Were this not so, there would after all be very little we could do but accept with resignation the fate of future generations. Hope rests in the very fact that the most direct threat is to something man has himself helped to shape in the course of history. What we need most urgently to protect is itself to a significant degree an artifact.

The basic components of our habitat – its geological structure and its envelope of climate – are less immediately under threat precisely because they are the aspects of the environment on which human activities have so far made the least impact. It is the biosphere itself – the vegetation and all the various forms of animal life – that is most directly at risk and it is precisely this that man himself has done most to shape. If we are to go about the task of conserving this most vital aspect of our environment intelligently and with discrimination we have to accept that it is to a significant degree culturally conditioned, the product of adjustments to the social needs of hundreds of generations of men. We are concerned not so much with a natural as with a cultural landscape.

Grahame Clark has been Disney Professor of Archaeology at the University of Cambridge since 1952, and has written widely on many aspects of European and global prehistory. He is a Fellow of the British Academy and holds the Hodgkins Medal of the Smithsonian Institution and the Viking Fund Medal.

The most complete and sensitive record we have of the extent to which men have modified their environment in the course of satisfying their economic needs is that provided by the detailed history of vegetational change during the Quaternary period. In reconstructing such changes, the technique of pollen analysis is all-important. It is a matter of unearthing and examining pollen of many thousands of years in age and deducing past patterns of vegetation from the counts of various species. The technique was first systematized between 1916 and the mid-1920s by Lennart von Post and his pupils and collaborators in Scandinavia; since then it has been applied in many other parts of the world and, in some areas, to a more extended range of time than that represented in the fossil soils of Sweden. As Sir Harry Godwin (1956) showed in his *History of the British Flora*, an early use of the method was chrono-logical, making it possible to assign to a time-zone any sequence of deposits that contained adequate pollen, and thereby to date any implied environmental change, or any archaeological indications of human activities. The post-war advent of radiocarbon analysis, allowing greater precision in dating deposits and the fossils they contain, has lessened the emphasis on the purely chronological value of pollen analysis. On the other hand it has only strengthened the relevance of its ecological concerns.

Magnus Fries (1965) reminded us, in his wide-ranging review of the study of the Late Quaternary vegetation of Sweden, that pollen analysis was first taken up and pursued with enthusiasm as a means of checking and amplifying hypotheses about climatic change originally formulated on the basis of grosser evidence. It is hardly surprising that the changes in vegetation revealed by pollen analysis and other studies should, from the beginning, have been interpreted as the outcome primarily of climatic history. Other environmental factors, such as changes of land and sea levels, were allowed a minor role.

The general picture, which still holds in its main essentials, under-lines the potency of climate. In Europe, towards the end of the last ice age ten thousand years ago, there were fluctuations between open vegetation and forest; then the forests spread rapidly during the post-glacial period, with the progressive rise to predominance of warmth-demanding species of trees as temperatures reached their peak. A slight reversion in northerly latitudes marked a decline in temperatures, towards the close of the pre-historic period. An ecological model in which there was no special place for man and his activities, and in which everything could be accounted for in terms of the interaction of purely natural variables, appeared to explain adequately everything revealed by early pollen analyses focused mainly on forest trees.

Not until the method had been refined and extended to include a much broader range of vegetation did anomalies begin to appear, including a peculiar episode of birch to which I shall return later. A decisive turning-point was publication of the late Johs. Iversen's classic monograph 'Land Occupation in Denmark's Stone Age' (1941). He had observed a number of interconnected changes in samples of the same age at various localities in Denmark. With elegance and penetration, he demonstrated that they denoted the use of land for farming. Iversen began his monograph, which was to set a whole generation of investigators to work in different parts of the world, by contrasting the situation in Denmark before and after the introduction of farming. He rightly saw it as a marked intensification of man's impact on natural vegetation. Before this happened, vegetational changes merely reflected changes in climate and other natural factors. 'All this,' Iversen wrote, 'went through a radical change with the introduction of farmer culture into the country. Then, and not before, the forest picture began to be altered by man, and there can be no doubt that the changes thus brought about were just as profound and just as interesting as those which earlier were associated with climatic and other purely natural causes. It would therefore be anticipated *a priori* that they would be clearly identifiable in the pollen diagrams.'

Although challenged on particular points, Iversen's general hypothesis has found wide support among colleagues in many parts of temperate Europe. His general theme was well illustrated by Frank Mitchell's study (1965) of Littleton Bog in Ireland. So long as man's numbers were small and his technology and degree of social integration remained at a primitive level, he had only a limited and, as a rule, locally restricted and temporary impact on vegetation. In Iversen's graphic words, 'Man's influence on the vegetation did not extend far; the virgin forest closed in a few paces outside the settlement, and there nature alone determined what was to grow'.

Even so, the influence of hunter-gatherers on vegetation was not quite negligible. In particular the use of fire in hunting and for making clearings must, at least locally and in the short term, have been a factor in vegetational change. Richard West's findings (1954) at Hoxne in England suggest that this was already happening a quarter of a million years ago. Pollen samples from a deposit contemporary with an encampment of Lower Palaeolithic hand-axe makers showed a marked decline in forest trees and an increase in grasses and herbs – something most easily explained in terms of local woodland clearance; the discovery of charcoal suggested that fire may well have been used for this purpose. Again, although the rapid spread of the shrub hazel in the post-glacial landscape over much of temperate Europe was made possible by the rise of temperature, it may have been encouraged by the inroads made on forest trees by mesolithic bands making limited clearings for their encampments and perhaps firing more extensive tracts in the course of hunting.

If pollen analysis has done nothing else, it has at least helped to demonstrate that early in the post-glacial period, even before Britain had been insulated from the Continent by the flooding of the North Sea, the land was already covered by forest. An airman flying over Britain in the eighth millennium BC might here and there have detected occasional plumes of smoke issuing from camp sites or more widespread forest fires, but otherwise his gaze would have been met by great tracts of forest, broken only by the upper ranges of mountain chains, by forest glades, and by rivers, lakes, marshes and coastal lagoons. Cultivated fields and meadows would quite literally have to be carved out of the forest, just as in much later times British pioneers were to clear tracts for farming in the eastern parts of North America (see Clark 1945).

So the fact remains that, as Iversen stated, the process of deforestation, which more than anything else was responsible for the emergence of the cultural landscape, was enormously accelerated by the adoption of farming. According to radiocarbon dating, this apparently took place as much as five or six thousand years ago, in north-western Europe.

The coming of the farmers

During the prehistoric period in temperate Europe, two main phases can be recognized in the development of farming. At first agriculture was of a shifting variety. Comparatively small plots would be cleared by felling and burning, the seed being raked among the ashes without the need to cultivate. Only two or three crops were harvested before new plots were cleared and the old ones were allowed to revert to forest. In other words the earliest farming was carried on by the slash-and-burn system, also known to have been practised in many parts of Africa, south-east Asia and the New World and which was still being carried on in eastern Finland in the twentieth century.

Pollen evidence for temporary clearance and cultivation has been found over a broad tract of territory from Ireland to Sweden. Samples from Neolithic levels show a decline of forest trees followed

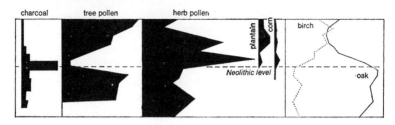

Fluctuations in the proportions of tree and herb pollen resulting from temporary clearance of forest by Neolithic farmers, Ordrup Mose, Denmark (after Iversen)

The charcoal peak argues that fire was used as well as the axe. The potash resulting from this would allow crops to be raised for a few years without the need to plough

by an immediate increase in shrubs, herbs, weeds and grasses, often including cultivated cereals; sometimes these finds are accompanied by charcoal (see diagram above). No less pervasive has been evidence for the abandonment of plots and their reversion to forest. In the process of regeneration birch trees, which flourish particularly well on ashy soil, were pioneers, only to decline again rather rapidly as their competitors spread.

This temporary maximum of birch, for which no evident climatic explanation was available, was what first alerted Iversen to the probability of human interference and opened up this whole new field for research. Experiments by the Danish National Museum on a hectare of woodland, using only fire, flint axes, seeds of einkorn, emmer wheat and naked barley, and branches for raking them in, have shown that it is quite practicable to grow crops without cultivation; and observation of the process of regeneration has confirmed Iversen's basic hypothesis.

Shifting cultivation must inevitably have come to an end when pressure of population on the lighter soils outstripped the capacity of the forest to regenerate. The transition to cultivation of permanent fields is likely to have happened at different times according to local circumstances. A useful clue to the appearance of the new regime is provided by traces in the subsoil of the actual marks of the light plough or ard as it was drawn across the ground first in one direction, then in another. The first man to recognize such markings was A.E.van Giffen of Groningen, who noted them in the old soil under an Iron-Age burial mound at Rhee in Holland. Similar markings from the Bronze Age and Iron Age were found in Denmark while, in England, confirmation that they in fact related to early cultivation came with their discovery in a visible Celtic field on Overton Down, Wiltshire. Further work at nearby Avebury suggested that in this part of Wessex at least the soil was being cultivated far back in the Neolithic period. The dates at which fields began to be cultivated in the various parts of Europe will only be determined by further research.

Meanwhile quite different evidence exists to document the advancement of agriculture during the course of the Bronze Age. In Holland, burial mounds from this time were generally sited on well-developed podsols and in some cases were actually constructed of turf sods, both features being consistent with a deforested landscape. Analyses of the pollen taken from samples removed from old ground surfaces and turf sods have fully confirmed this interpretation and show that the landscape was already deforested when many of the mounds were constructed. Comparable results have been obtained from different parts of England. For instance G.W. Dimbleby (1954) was able to arrange a series of barrows from Dorset and Hampshire in chronological sequence on the basis of the degree of deforestation they exhibit.

Even more unambiguous evidence for cultivation is provided by the so-called 'Celtic' fields of southern England. These are generally short, oblong in shape and are defined by boundary strips formed by a certain degree of terracing on hill-slopes. Most of these were still being worked by British peasants in Roman times, but some of them were in use well before the end of the Bronze Age. These ancient fields survive beyond the normal modern margin of cultivation and it is likely that many more, including perhaps the earliest ones, have since been ploughed out of existence.

73

Our aviator, if he could have flown over Britain during the closing phases of the prehistoric period, would have seen extensive tracts of open country, more particularly on the gravel terraces of the great rivers, on the chalk downs and limestone hills, on wolds, moors and brecks. If he were a good observer, he would have spotted cultivated fields among the pasture, at any rate in the south of England. He would nevertheless have seen extensive tracts of more or less virgin forest.

The process of clearing the forest and converting it into grazing and arable land was a lengthy one. As the late Sir Cyril Fox (1923) first showed, farming was confined during the prehistoric period in southern Britain to the lighter, better drained and more easily cultivated soils, notably gravels, sands, chalk and limestone. The heavier claylands, potentially richer but more difficult to cultivate, were not taken in until much later. The Romans drove their roads across them but they seem first to have been taken into husbandry to any substantial extent by the Anglo-Saxons.

Evidence for the early stages in the clearance of the secondary area of settlement on the clay soils is scanty. As Hunter Blair (1956) has emphasized so well, very little is known about the progress of clearance by the Anglo-Saxons, even though we may be sure that they contributed in decisive fashion to creating 'the open face of the English countryside and its place-names'. By the eleventh century AD, the distribution of settlement recorded in the Domesday Book shows a marked shift to the heavier claylands. Indeed if one compares the Domesday pattern with that of modern settlements, around Cambridge for example, a remarkably close measure of agreement appears in this particular region. The basic pattern of English settlement was already defined before the close of the Anglo-Saxon period, even if in areas like the wooded Chilterns or parts of Essex there was room for many new settlements during the later Middle Ages (see Hoskins 1957). In the same way there are indications, not yet very precise, that the Anglo-Saxons made some modest progress in the task of reclaiming marshlands, for example the Pevensey Levels in Kent.

The extensive tracts of forest that remained were by no means waste areas. They formed an essential part of the rural economy, providing materials for buildings, timber for a variety of crafts, fuel, shelter for game and pannage for swine, to mention only some of their more important uses. Similarly, as the commissioners

appointed in connection with later drainage schemes were to discover, the marshes and fens were also sources of wealth, notably in
the form of fish and fowl which their denizens were loth to lose even
in the cause of progress.

The modern landscape

Although by the time of the Norman Conquest the English landscape may already have acquired its 'open face', the precise features
of this face were very different from those with which we are
familiar. The pattern of the present landscape is to a large extent the
product of comparatively modern history. The process of enclosing
the open fields of the medieval village and abolishing the multitude
of narrow strips began in the fourteenth century and was carried
forward at an increasing tempo during the centuries immediately
following. It was only one aspect of the transformation of a society
based on hereditary obligation, to one held together by a nexus of
cash and deriving its wealth increasingly from manufacture and
trade as the population began concentrating in towns. The growth of
the textile industry, for instance, made it more profitable to raise
sheep than crops and in many parts of the country this led to the
contraction of villages and often to their abandonment.

The final assault on the forests was mounted in a similar context.
Thus the fate of the Wealden Forest was determined by the need for
charcoal for smelting iron. The Romans made a beginning, as we
are reminded by their use of iron slag for metalling their road from
London to the Brighton area, but the Sussex ironmasters in more
recent times were those who finally taxed the woodlands beyond
their capacity to regenerate. Similarly, although the Romans and
even the Anglo-Saxons and the later medieval monks made tentative
inroads on fens and marshes, great enterprises of drainage, like those
mounted in the East Anglian Fenland, had to wait on the markets
and capital provided by the industrial age. Even the extensive land-
scaping associated with the great country houses – the avenues,
parks and coverts – was paid for from the profits of capitalist enterprise, carried on by the landed aristocracy at least as much as by new
men from the cities.

By the time the modern cities, with their suburbs and their attendant sewage works and reservoirs, not to mention the communications that knit them together and the power plants and factories

75

that sustain them, first spread over the countryside they were not invading a natural, but an artificially shaped landscape. Indeed, the very structures of industrial society can be regarded as inescapable elements in the cultural landscape of modern man. It is hardly logical to approach environmental problems with a naive wish to protect nature from man, when so much that appears natural is in fact a social artifact. The extent to which existing vegetation has been influenced by man is shown by Ritchie's diagram for Scotland.

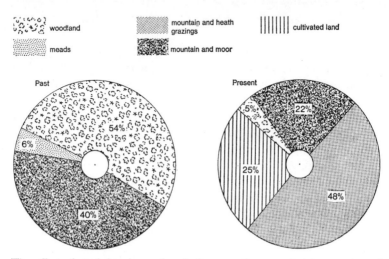

The effect of stock keeping and agriculture on the post-glacial vegetation of Scotland (after Ritchie 1920)

When it was first proposed to plant forest trees on the Norfolk breckland after the First World War, many of those who wrote furious letters to the newspapers seemed to imagine that they were concerned with a primeval landscape. Pollen analysis (Godwin 1944) has since proved that the breckland vegetation was in fact a human product. It is the work of Neolithic man and his successors who between them cleared away the deciduous forest that had itself replaced the open landscape of the late glacial period.

One moral seems to be that in crowded countries like Britain we should seek to preserve, not some mythical 'natural' environment, but examples of the many cultural landscapes that came into existence during different phases of our social history. Part of the mental blockage which inhibits the kind of solution we need is enshrined in the title of such a body as the Natural Environment Research Council in Britain or, again, in the restricted scope of the United Nations Conference on the Human Environment at Stockholm in 1972. As Lili Kaelas (1972) commented:

> . . . the concept of environment is wider than what was scheduled for the UN Conference. It comprises much more than nature. Our physical surroundings are mainly a cultural product and this is true about the agricultural landscape and the forests as well as the cities and the industrial landscapes and also about a large part of seemingly virgin nature.

Concentration on the conservation of features or localities of natural beauty or cultural interest as though these are separate entities is no longer sufficient. The more comprehensive scope of the new Department of the Environment in Britain is a step in the right direction. If we are adequately to safeguard our heritage, as manifest in our cultural landscape, ancient monuments like those on Dartmoor or Salisbury Plain need to be treated not as isolated features but as integral parts of landscapes in the manner pioneered by the United States National Parks Service. As this chapter has done no more than outline, relic environments provide irreplaceable archives of man's long experience as an engineer of the environment.

References and further reading

Peter Hunter Blair, *An Introduction to Anglo-Saxon England* (Cambridge 1956).

J. G. D. Clark, 'Farmers and forests in Neolithic Europe', *Antiquity*, 29 (1945), pp. 57–71.

J. G. D. Clark, *Prehistoric Europe: The Economic Basis* (London 1952 and 1965, New York 1966).

J. G. D. Clark, *Excavations at Star Carr* (Cambridge 1954 and 1971).

G. W. Dimbleby, 'Pollen analysis as an aid to the dating of prehistoric monuments', *Proceedings of the Prehistoric Society*, 20 (1954), pp. 231–6.

Knut Faegri and Johs. Iversen, *Text-book of Modern Pollen Analysis* (Copenhagen 1950).

C. Fox, *The Archaeology of the Cambridge Region* (Cambridge 1923 and 1948).

Magnus Fries, 'The late-Quaternary vegetation of Sweden', *Acta Phytogeographica Suecica 50: The Plant Cover of Sweden* (Uppsala 1965).

H. Godwin, 'Age and origin of the "Breckland" heaths of East Anglia', *Nature*, 154 (1944), p. 6.

H. Godwin, *History of the British Flora* (Cambridge 1956).

W. G. Hoskins, *The Making of the English Landscape* (London 1957).

Johs. Iversen, 'Land occupation in Denmark's Stone Age', *Danmarks Geologiske Undersogelse* II.R., 66 (Copenhagen 1941).

Lili Kaelas, 'Museum – man – environment', *Helinium*, 12 (1972), pp. 117–38.

G. F. Mitchell, 'Littleton Bog, Tipperary: an Irish vegetational record', *The Geological Society of America Special Paper* 84 (INQUA USA, 1965).

James Ritchie, *The Influence of Man on Animal Life in Scotland: a Study of Faunal Evolution* (Cambridge 1920).

R. G. West and C. B. M. McBurney, 'The Quaternary deposits at Hoxne, Suffolk, and their archaeology', *Proceedings of the Prehistoric Society*, 20 (1954), pp. 131–54. See especially p. 135.

Industrial geography 8

THE DISAPPEARANCE OF MOUNT FUJI
Bryan E. Coates

In its early phases, human industrial activity creates recognizable types of community, in which particular occupations, sights, sounds and smells are dominant: 'mining towns', 'steel towns', 'chemical towns' and so on. Zones of greater size and more diversified in composition have now entered the world's industrial scenery. With large populations and all kinds of ancillary services for homes and factories, such industrial complexes are like an amalgam of scores of towns of the older sort. The more the industrial system reflects man's growing sophistication in moulding the resources of the natural and human environments, the more it comes to resemble, in its intricacy, the ecological systems studied by the biologist. As one moves away from the traditional steel or chemical town towards the industrial complexes of, say, the Tokyo Capital Region or south-east England, the industrial 'ecosystem' becomes more difficult to unravel and comprehend, or to manage for the benefit of its inhabitants.

The growth of industrial complexes

Industrial towns of any kind are a very recent phenomenon. Even in advanced industrial countries the factory, as opposed to the

Bryan E. Coates is a senior lecturer in the Department of Geography at the University of Sheffield. Apart from his recent studies of industrial development in Japan, his research has been largely concerned with economic and social variations between the regions of Britain.

cottage workshop, has dominated manufacture for less than two centuries. Although it might be argued that northern Italy and Flanders were industrial areas as early as the fourteenth century, modern industry developed first in Britain. It was characterized by machines driven by inanimate energy and by large, specialist operating units. Concentration on a limited range of products with specially trained workforces increased productivity and lowered costs.

During the nineteenth century, industrial towns sprang up in many parts of Western Europe and North America. The steamship and the railway gave industrial man, in the age of steam and steel, the ability to organize almost the whole world for the benefit of a few, industrially advancing nations. Within the latter, inventions of 'practical man' and 'scientific man' produced the age of electricity, oil and the internal-combustion engine. Each major breakthrough called for new resources and products, better methods of organization, new transportation systems and additional sources of energy, more capital investment, more sophisticated financial services and more extensive trading links. As 'know-how' multiplied, it became difficult for a new industrial area to develop without help from an older one. In the nineteenth century, European engineers sallied forth to build ports, create railways, install new plant, train labour and generally share the secrets of the new way of life. Up to the present, individual members of the industrially advanced 'club' have vied with each other to provide the technical base for new industrial societies abroad.

Yet, even now, industrial towns and complexes occupy a very small part of the Earth's surface. Unlike farming, which tends to form continuous belts sufficiently extensive to appear prominently on world maps, manufacturing requires relatively little space even when it is clustered in major agglomerations. The latter are so far confined to four areas – the manufacturing regions of Europe, east-central North America, the Soviet Union, and eastern and southern Asia. Smaller concentrations of industry can be found in every continent.

Despite its compactness, industry produces far-reaching geographic effects. Industrialization has led to a huge redistribution of economic activity. Previously unvalued mineral resources, especially coal and later oil, became as important as land, climate and biological resources. Minerals are generally concentrated in small areas, so industrial towns originated either on the spot or in nearby areas where the resources could be tapped by means of efficient transport. As a result of the economic benefits of agglomeration, the favoured areas attracted not just one or two settlements but sizable clusters, for example of mining towns, textile towns, chemical towns or car-assembly towns. Regional specialization in a given industry or group of industries emerged as a characteristic feature of an intermediate stage of industrialization, still discernible in the Lancashire textile and Californian aerospace industries. But as industry diversifies and becomes increasingly interdependent, and as population becomes concentrated, the economies of proximity and scale and the attractions of the large market favour the development of huge, intermingled industrial conglomerations producing a great range of products.

Such industrial complexes are often centred on a pre-existing metropolitan area and, as they grow, they sharpen the contrast between the prosperity of the cities and the relative poverty of agricultural areas. The older industrial areas, too, often located on coalfields, fall behind because the market-oriented industries, and new growth industries in general, are drawn to the metropolis as if by a magnet, leaving the old 'staple' industries in their nineteenth-century settings. As particular resources are exhausted, the search is extended for new supplies and the tentacles of industrialization spread farther afield. The industrial complexes remain relatively small but the area affected by their voracious appetite continues to expand.

This consumption of innumerable raw materials by innumerable processes on a huge scale in industrial complexes, occupying relatively small parts of the Earth's surface but with large populations, produces its own inimitable environmental effects. No other nation in modern times has managed to make quite such a mess of its environment as have the Japanese, as they now officially admit. But, in this respect, Japan differs from other industrial countries only in degree and, as the extreme example, it shows most clearly the consequences of environmental mismanagement of which many countries have been guilty.

The disappearance of Mount Fuji

With a land area of only 370,000 square kilometres, Japan is by far the most advanced industrial country in Asia. Its population has grown from 70 to 100 million since 1945. By 1965 less than one-quarter of its workforce of almost 50 million was employed in primary occupations and almost one-third was employed in the manufacturing sector. Industrialization has been so rapid and the development of many different forms of pollution so severe that the Japanese have coined the term *kōgai* to refer to man-made environmental hazards. It means 'public hazard' and in the Basic Law for Environmental Control, promulgated in 1967, it is defined as the harmful disruption of human health and the living environment by pollution of air and water, excessive noise, vibrations, land subsidence and offensive odours. The living environment is defined as including property, animal and vegetable life necessary for the support of human existence, and the environment necessary for their growth and development.

More newsprint is probably devoted to pollution in Japan than in any other country. As one observer has remarked,

Japan may also be the country of the highest ratio of dire predictions to concrete prevention measures. There is a daily plethora of print, a whirlwind of words prophesying the extinction of life on these islands in fifty years while the despoliation of the environment goes on; Osaka's citizens continue to cough their way to work (the city's Nishinari ward lists half of its residents as sufferers from asthma brought on by soot, smoke, and sulphur dioxide emitted by factory chimneys and car exhausts) and Tokyoites watch anxiously as their city's trees choke in photochemical smog (Whymant 1972).

The more dramatic tragedies have been reported in the rest of the world's press: the victims of Minamata disease who ate fish contaminated by industrial waste containing methyl mercury compounds; the sufferers from *itai-itai* ('it hurts – it hurts') poisoned by cadmium wastes; the officially designated victims of water pollution and of air pollution; and so on.

The view of Mount Fuji from Tokyo, once taken for granted as a source of pride and inspiration, has been all but obliterated by a perennial smog that hangs over the world's largest city. As one speeds close by the mountain on the incomparable Tokaido Express it makes a backcloth to a plain of appalling squalor, part covered by Fuji City. Paper mills, chemical and food-processing plants, chimneys belching smoke, an indiscriminate scatter of low-density houses, danchi (public authority apartments), neglected open sites and a confusion of poles and overhead wires stand in tribute to the thrust of opportunism and to careless regard for community values. The contrast with the beauty of the snow-capped mountain is indescribable (James 1972).

Why do the industrial complexes of Japan generate such severe environmental problems? Why is the air so polluted, the rivers, lakes and ocean margins so contaminated, the noise levels so disturbing, the land subsidence so marked, the environmental destruction of such an inherently beautiful land so appalling? Several reasons conspire to produce these results.

RAPID GROWTH. The major cause is undoubtedly the extreme speed of industrialization of recent years. Although Tokyo itself first experienced rapid industrial development at the time of the First World War (between 1911 and 1921 its industrial workforce more than doubled) the most dramatic industrial expansion occurred after 1955. Since then Japan has had an annual economic growth-rate far in excess of anything ever experienced in the Western World. Between 1966 and 1970 alone the nation's gross national product doubled and Japan ranked third in the world league table of production, behind only the United States and the Soviet Union.

Among the remarkable increases in output between 1955 and 1969 were those in electric power (a five-fold increase), crude steel ($\times$ 9), synthetic-fibre fabrics ($\times$ 44), television sets ($\times$ 93), refrigerators ($\times$ 100), and automobiles ($\times$ 126). The most environmentally alert government would be hard put to plan for the absorption of such a huge growth in industry.

By 1967, five million metric tons of carbon monoxide was being discharged annually into the atmosphere – one-quarter of it in the two cities of Tokyo and Osaka – and before long Tokyo was experiencing photochemical smogs. In August 1970, after a serious occurrence of photochemical smog accompanied by a sulphuric-acid mist, the Tokyo Metropolitan Government began issuing a photochemical air-pollution forecast. Industrial pollution is particularly severe in cities like Ichihara, Kawasaki, Yokkaichi and Kita

Kyushu with their petrochemical complexes, iron and steel industries, cement factories, thermal power stations and heavy chemical industries.

CONCENTRATION OF INDUSTRY. Environmental damage is heightened because industrial activity is by no means evenly spread in Japan. Pollution is concentrated in particular areas because Japanese industry is itself concentrated in a few gigantic industrial

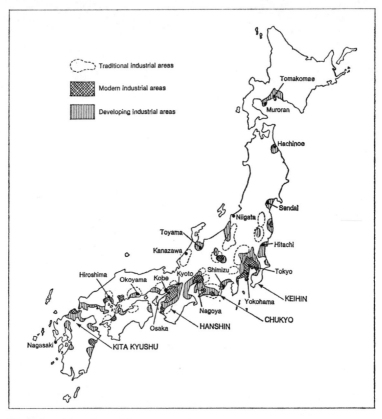

The industrial regions of Japan

complexes. The Pacific coastside of Honshu, and especially the Tokaido belt from Tokyo to Kobe, has taken the brunt of the industrialization of the post-war period (see map). The three great complexes – Keihin (Tokyo-Yokohama), Hanshin (Osaka-Kobe), and Chukyo (Nagoya) – employ more than half the labour force in every industrial group except that of wood and pulp. Outside the industrial towns and complexes, many Japanese still live an almost traditional life.

Despite some tokens of dispersal from the most congested areas, Japanese industry has become ever more highly concentrated. The centre of gravity of the main industrial belt has shifted eastward because of the tremendous growth of the Tokyo area, increasingly spilling around Tokyo Bay into the southern Kanto Plain. Here, in Saitama, Chiba and Kanagawa, the future is bleak in terms of *kōgai*, though not in terms of new jobs and civic wealth. Industrial concentration per unit area in the Kanto Plain is still far below that of the Osaka-Kobe area, for example, and the potential for further industrial development is, by Japanese standards, enormous. The government planners foresee sustained expansion spreading outwards from Tokyo for some 100 to 150 kilometres and this one area containing 55 million people (the present population of the UK) by the year 2000. The attempts of the Japanese government to disperse industry away from the gigantic complexes have so far been half-hearted.

CONCENTRATION OF PEOPLE. A third major cause of the intensity of environmental disruption is that the majority of the Japanese people occupy a very small proportion of the country. A high rate of migration from rural areas into the cities has been one of the by-products of the industrial boom. This, in turn, has accentuated the regional imbalance and also the environmental pressures within the major centres of population.

Although most people now live in cities, the entire built-up area occupies only 2 per cent of the land surface. Well over half the population resides in the three main metropolitan areas of Tokyo, Osaka and Nagoya. Even more dramatic evidence of the degree to which the population is concentrated is seen in the official data on Densely Inhabited Districts. These are areas containing 4,000 people or more per square kilometre (more than 10,000 per square mile, or about the same as Hong Kong and Singapore). In 1960 the DIDs covered 1.05 per cent of Japan and included 40.8 million people; five years later the respective figures were 1.25 and 47.3.

In a country with as little flat land as Japan, industry competes with human beings for space to grow. It is estimated that the gross national product per square kilometre of level land is eleven times that of the United States. The concentration of waste products is correspondingly high. Such intensity of land use both fosters environmental disruption and hinders attempts to check the inexorable rise in pollution levels.

SMALL-SCALE ENTERPRISE. Japan is an industrial country dominated by small workshops rather than large factories. Its industrial structure makes it very difficult indeed for local and central government to control environmental damage. By 1967 there were 83,820 factories in Tokyo alone and 70 per cent of them had ten employees or less. Two Tokyo wards accommodated 600 factories per square kilometre. Workshops are often embedded within the residential and commercial areas and their owners have scant reserves for financing anti-pollution measures. The flimsy construction of many workshops is partly responsible for the severe noise problem in Japanese cities. In the whole of Japan in 1969 there were almost 650,000 industrial enterprises and the table below gives some indication of their size and importance, as measured by employment.

Industrial enterprises in Japan 1969

Number of employees by industrial enterprise	Number of establishments	Percentage of establishments	Percentage of workforce
under 10	475,058	73.4	16.6
10–99	156,597	24.2	35.6
100–999	14,393	2.2	30.4
1000 and over	878	0.2	17.4

The problem of controlling so many enterprises would be difficult for a country with an effective body of legislation regarding pollution and the necessary will to enforce it, with a modest economic growth rate, with stable firms and with powers for regulating the use of land. None of these conditions exist in Japan. The Western visitor is immediately struck by the absence of zoning and by the 'indiscriminate scatter' of homes and factories. Although more rigorous control of land use is now being attempted, the recent lack of effective zoning laws to separate functions has resulted in chaos.

Moreover, the unceasing urban and industrial pressure has spilled over from the main centres into contiguous areas. Even in these 'new' areas the opportunity to exercise some control over land use for development has not been taken. During the past two decades of unusually rapid industrial growth, both zoning and anti-pollution regulations have been far too weak to deal effectively with the tremendous pressures unleashed – for land, for water, for air, for effluent disposal, for transport systems and so on – by the multitude of busy Japanese entrepreneurs.

NEGLECT OF PUBLIC SERVICES. The rapid aggregation of industry and population into the belt of great urban concentrations is the result of the injection of private and public capital. This investment has not been matched by investment in public utilities or in the equipment and development of the urban infrastructure in general. Many of the environmental nuisances are the result of unrestrained concentration of industrial activity within the already urbanized areas of the country. Public expenditure even on water supplies and the treatment of sewage is inadequate to keep up with basic needs in these areas, let alone the treatment of industrial waste, the creation of open spaces or the development of welfare schemes. The last are notably lacking for the poorest citizens, who tend to live in the most heavily polluted parts of the major cities.

Confessions of government

In 1972 a popular Tokyo newspaper asked, 'What has the Sato Administration left behind ?' and supplied its own answer: 'A dirty environment in the name of economic growth'. Yet Mr Sato's government did make the following public statement which cogently expresses the fundamental conflict that now exists in Japan, between economic growth and quality of life.

Too eager to raise our living standards and too anxious to catch up with the material prosperity of Western nations, most Japanese leaders have been the captives of the disease of growthmanship, even after such attitudes became unwarranted. Businesses, with a few exceptions, have paid little attention to what their activities would do to environment. Scholars and journalists have not realized the deterioration of environment or, at any rate, have not tried hard enough to

87

demand preventive measures. The Government has spent too large a proportion of public funds for productive investment, neglecting social services. It has failed to make adequate zoning plans to segregate the residential from industrial areas. It has done little to regulate the activities of polluting businesses except when the harm was apparent. In the meantime, environment disruption has proceeded in a creeping way without drawing much attention at first, but then, with the quickening pace of technical progress and urbanization in the post-war years, it has developed into monstrosity.

The Japanese are rightly proud of their economic achievements over the past two decades. A high premium has been placed on massive capital investment for future growth. Government and big business are closely linked and both realize that more effective measures to bring pollution under control would mean diverting resources to 'non-productive' uses, thus lowering the growth rate. Many Japanese economists hold the view that an annual growth rate of about 8 per cent will be required throughout the 1970s if resources are to be freed for massive investment in socially desirable projects, which threatens a vicious circle of increasing pollution to cure pollution.

In the meantime, Japan is the prime example in the modern industrial world of what happens in a country which undergoes very rapid economic growth, which is prepared to make great sacrifices in the pursuit of material wealth and which allows its industrialists almost a free hand to produce their goods without having to give much thought to the social costs of their activities. The environmental consequences are a warning to other nations. Japan has shown that, even in the second half of the twentieth century, it is still possible to create a complex of 'dark satanic mills' in which man is subjugated to production in a rapidly worsening environment. It remains to be seen whether a 'cowboy economy' will give way to a 'spaceship economy' in which the quality of life will stop deteriorating and begin to improve.

References and further reading

C. A. Fisher, 'The maritime fringe of East Asia: Japan Korea and Taiwan', *The Changing Map of Asia*, ed. W. Gordon East, O.H.K. Spate and C. A. Fisher, 5th ed. (London 1971).

J. R. James, 'Japan', *Guardian* (28 June 1972).

Japan Development Bank, *Statistical Analysis of Economic Growth, by Region, in Japan* (Tokyo 1969).

T. Kawashima, 'Some trends and problems on the recent regional development of manufacturing industry in Japan', *Economic Review, Osaka City University*, 2 (1966), pp. 33–41.

Ministry of International Trade and Industry, *Present State of Regional Development in Japan* (Tokyo 1970).

Y. Okuda, 'Economic policy and regional development in Japan', *Bulletin, Faculty of Science and Engineering, Chu-o University*, 12 (1969), pp. 174–94.

Tokyo Metropolitan Government, *Tokyo Fights Pollution* (Tokyo 1971).

R. Whymant, 'Civilisation going up in smoke', *Guardian* (7 July 1972).

HUMAN
SETTLEMENTS
IN SPACE
AND TIME
C. A. Doxiadis

Many urban regions today are scenes of evident instability, with change and decay creating human and environmental problems that planners and municipal administrators sometimes despair of solving. By contrast, archaeologists know of villages that survived for thousands of years, and historians of city-states that prospered and preserved their character for many centuries. Thus our ancestors, by trial and error, created villages where their actions were in balance with the demands of the whole system of nature and human life; similarly with city-states in many ancient civilizations. After a long period of neglect, this system of life is again beginning to be of interest to man.

It might be argued that the interests which villages and city-states had in balance were only incidental, and that their success was based on unconscious efforts. Yet, in most of the historical cases we know, man did in fact achieve a balance and in several of them, at least, he was fully aware of what he was doing. When Aristotle, in his *Politics*, wrote about Hippodamos 'who invented the division of

C. A. Doxiadis has planned the development of human settlements in many parts of the world, and was the post-war minister of reconstruction in Greece. He is now president of the Athens Center of Ekistics and of Doxiadis Associates International.

cities into blocks and cut up Piraeus' he did not simply mean urban planning in our modern sense; he also meant awareness of social and political aspects of city life as well as of more widespread physical factors such as requirements for farming land. Plato showed a similar attitude in suggesting that each citizen should be given both urban and rural land.

I use the term 'physical planning' to include the planning of towns, cities, regions and even the whole world, from the point of view of architecture, engineering and all the other technological and physical components of modern human settlements. But we now have to master not only the conventional disciplines of physical planning but the real subject of our concern, human settlements seen in the broadest context. Human settlements are the cause of environmental problems because without them nature would have followed its own course; they are also our hope for the future because of the historical evidence that many cultures and civilizations have successfully established balanced relationships between their settlements and nature.

Unfortunately, the disciplines related to physical planning today concentrate particularly on land use for new urban development. In doing so, they all make a basic mistake. They deal only with the area where current building is in progress – the 'invasion area'. This preoccupation means that they think about only very small parts of the areas where we live and where human action takes place, and they see human settlements from the point of view of man the builder rather than of man the producer, man the culture-seeker or man as a part of nature.

To correct the present bias among architects, engineers and planners, I have tried to develop a science of human settlements. *Ekistics* (Doxiadis 1968) seeks to study not only the area where we build but the whole system of our life, which is seen as consisting of five elements: nature, man, society, shells (or buildings) and networks, which may be roads, sewers or telecommunication systems.

The complexity of human settlements

The interactions of these five elements vary tremendously from space-unit to space-unit, whether village, town, large city, metropolis or 'Ecumenopolis', the emergent world city. From time-unit to time-unit too, the problems are very different; smoke bothers me today but the wind may remove it tomorrow, while pollution from certain chemicals may last for centuries. Any generalized statement about human settlements, or proposal for action, is likely to be naive unless it takes account of what is appropriate for each space-unit and time-unit. We also have to realize that our opinions about the components of human settlements depend on the criteria we use. A factory may create grievous chemical, biological or aesthetic problems but it can save a town economically by eliminating un-employment. So we have to reconcile what is desirable with what is feasible.

The units of space and time, together with the wide variety of criteria of choice, constitute a range of vantage points from which to view the five elements of human settlements. These elements, in turn, are divisible into constituent parts. 'Nature', for example, includes climate, land, water, air, flora and fauna, and mineral resources. Proceeding in this way, we can build a model of man's world (see opposite) which presents all aspects of human settlements. If we then consider the mutual dependencies of the elements, we find we are dealing with a system of interconnections of the order of one hundred million.

Such a model shows the tremendous complexity of human settlements. It may frighten those who have not dealt with the problem seriously, but this is reality. Unless we keep in mind all parts of such a model, we overlook so many aspects that it is impossible to make a success of physical planning.

With such a model it is easy to see why specialists dealing with cities so often accuse experts in other disciplines of failure; the bad architect blames the civil engineer for the defects of his buildings and the bad regional planner blames the economist when his plan is not implemented. The model shows, to all concerned, the immensity of the areas of relations between parts, of problems and naturally of present ignorance. It also lays the foundation for a strategy to break down mental barriers and connect all disciplines – not in the interdisciplinary anarchy which so often nowadays squanders its time in chaotic discussions, but in the building of a team which, guided by those who grasp the total system of human settlements, brings all the necessary hard-headed expertise together.

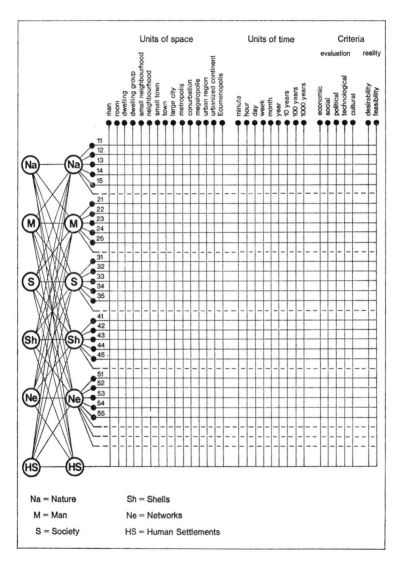

A schematic model of the world, as it relates to human settlements

The space-units

To show why we need a model like the one described, I shall extract one of its features, the range of units of space, and indicate the very great change of the system of life from one unit to another. I distinguish fifteen space-units, from man himself and the small traditional units to the whole Earth as it may be in the foreseeable future (Doxiadis 1968).

Ekistic unit	Population (foreseeable future)
1 man	1
2 room	2
3 dwelling	4
4 dwelling group	40
5 small neighbourhood	250
6 neighbourhood	1,500
7 small town	9,000
8 town	50,000
9 large city	300,000
10 metropolis	2 million
11 conurbation	14 million
12 megalopolis	100 million
13 urban region	700 million
14 urbanized continent	5,000 million
15 Ecumenopolis	30,000 million

When man builds these various space-units, many forces are at work. In the small unit of a room, man's anatomical and physiological needs lead us repeatedly towards certain solutions. These needs dwindle in relative importance in the larger space-units, where other forces such as topographical ones prevail. When we build a house, we can at reasonable cost change the topography of the little piece of land it stands on, but a megalopolis is constrained by coastlines and mountains. The diagram opposite shows, in a general way, how classes of problems vary from space-unit to space-unit and therefore demand different solutions.

Let us focus particularly on the conurbation (ekistic unit 11). This is really the space-unit which nowadays corresponds to what man has always been calling a city, because the city was defined by the extent of man's normal daily movements. We still tend to call 'city' the compact area defined by the movements of walking man,

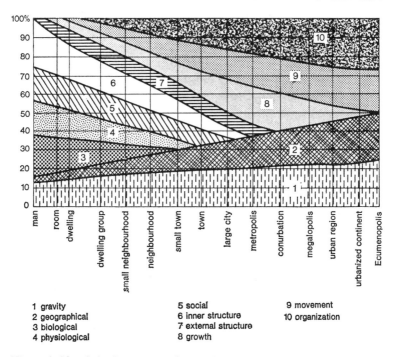

1 gravity   5 social   9 movement
2 geographical   6 inner structure   10 organization
3 biological   7 external structure
4 physiological   8 growth

The probable relative importance of the main constraints on human settlements of different sizes

which used to mean an urban radius of up to 5 kilometres. When the Mid-West was settled, the United States Congress decided that a township should be 10 × 10 kilometres. Today, man commutes over much larger areas and thus redefines the real dimensions of the contemporary city. The average American conurbation has a radius of 92 kilometres. The real city, the 'daily urban system', has grown about three hundred times in area since the time of Thomas Jefferson.

We have made studies of several such conurbations, the most detailed being for the Urban Detroit Area, the Northern Ohio Urban System (Cleveland's daily urban system), Catalonia (Barcelona's daily urban system), and Southern France (an urban system of several cities including Marseilles, Nice, Montpellier, Nîmes and Perpignan). They all have common characteristics, with a population of between five and eight million and a radius of up to 200 kilometres. They are the urban systems within which people

95

are beginning to commute; they are becoming typical 'real cities' for the next few generations. As the diagram on p. 95 indicates the problems of conurbations are largely those of organization, movement and growth.

From the dozens of aspects we have studied in every one of these urban systems, with a view to better physical planning, I select for illustration the case of networks around Detroit. This comes from a joint project of the Detroit Edison Company, Wayne State University and Doxiadis Associates. The first of the maps shows the existing maze of freeways, pipelines etc. and it is easy to see why these networks constitute the main environmental disasters in the countryside. If the networks were unified, into transportation and utility corridors, it would be possible to save great amounts of land at present occupied by the networks – perhaps more than 80 per cent. This would be accompanied, not only by a great reduction of the ill-effects on the natural environment, but also by great economies in the structure and operation for all the constituent networks.

The second map illustrates the most desirable and feasible arrangement of these corridors selected from a range of solutions, calculated to be, in principle, of the order of forty-nine million different possibilities.

Through co-ordination of networks we can preserve or improve the environment while providing excellent services represented by the networks. For example, when Detroit creates its transportation and utility corridors, environmental values represented by the small lakes north-west of Detroit can be saved as building follows other directions. The third map shows how the Detroit region should look thirty years from now.

Similar analyses, exploiting modern methodologies like IDEA (Isolation of Dimensions and Elimination of Alternatives), demonstrate how, in Catalonia, the Costa Brava can be relieved of increasing pressures; how the Mediterranean part of France can be unified; how the future transportation system around Cleveland can bring in life that Northern Ohio badly needs.

The future of ekistics

The methods and cases presented here are the results of hard, practical work in the development of a science of human settlements. Forty years of personal effort, nowadays multiplied by the

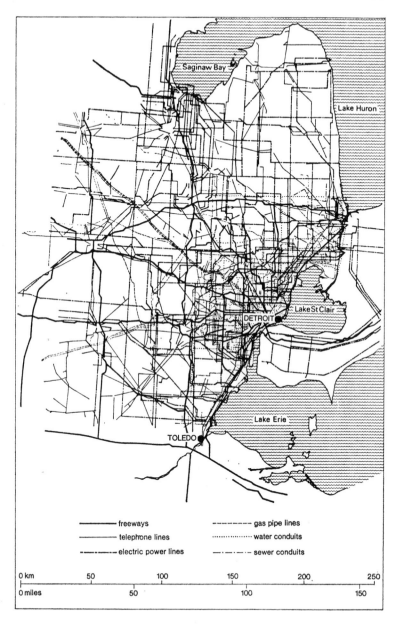

The networks of the Detroit 'daily urban system' as they existed in 1970

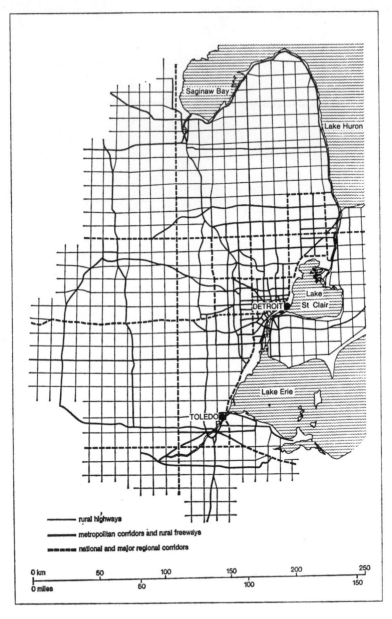

A possible system of co-ordinated networks for the same area

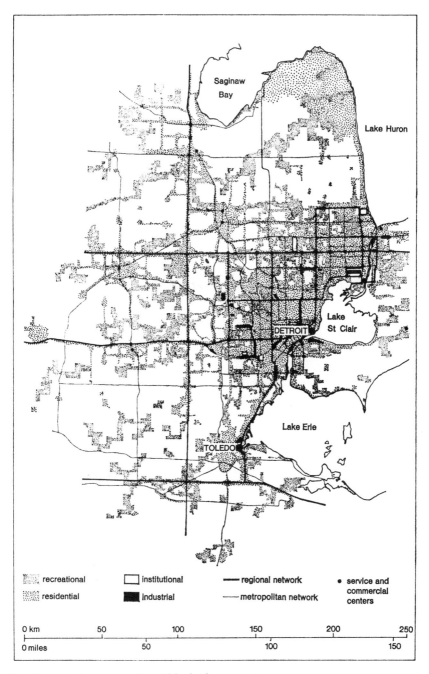

The Detroit urban area as it could be in the year 2000

work of our consulting firm and the Athens Center of Ekistics, convince me that human settlements have been governed by the same principles since the beginning of history. By finding these principles, in systematic study, we can project realistically into the future, understand our increasing environmental problems, and find solutions to them.

The foundation for such an effort is research based on measurements of the physical structure of human settlements in space and time. Only the physical structure can give us the frame of reference needed for understanding any phenomena related to the system of human life, whether biological, technical, aesthetic, social, political or psychological. For instance, patterns of crime in cities and in the countryside cannot be interpreted unless we measure proximity, and the exposure of people to each other, to name only two of many factors. A criminal in one place may be a criminal in the other, but his action depends on where he performs it and how it is connected with other people.

If we develop such a co-ordinated system of studies, we can encourage all the relevant disciplines to interconnect and thus help man to determine what he wants (which he does not yet know) and how he can achieve it, with harmony as the ultimate goal of the physical planner. Meanwhile, we must stop indulging in vague nightmares and dreams and get on with specific action in real cities.

References and further reading

Délégation à l'Aménagement et à l'Action Régionale, *La Façade Méditerranéenne* (Paris 1969).

C. A. Doxiadis, *Emergence and Growth of an Urban Region, the Developing Urban Detroit Area*, I (Detroit 1966), II (Detroit 1967), III (Detroit 1970).

C. A. Doxiadis, *Ekistics, an Introduction to the Science of Human Settlements* (London and New York 1968).

C. A. Doxiadis, 'The future of human settlements', *The Place of Value in a World of Facts* (Nobel Symposium 14), ed. Arne Tiselius and Sam Nilsson (Stockholm 1970).

C. A. Doxiadis, 'Ekistics, the science of human settlements', *Science* 170 (1970) pp. 393–404.

Ekistics, 1–34 (1957–72) and especially 33, No. 199 (June 1972).

DILEMMAS IN SAFEGUARDING HEALTH
W. Harding le Riche

Tuberculosis is still a serious cause of death and disability in all the poorer countries of the world, as it was until recently everywhere. It has been brought well under control in most of Europe and North America, where it is now relatively uncommon, although still a danger. When we look at the reasons for this success, in the wealthy parts of the world, we see it has depended upon a whole group of situations which include isolation of infective cases and novel pharmaceuticals – but also better nutrition, less crowding in housing, less physical strain at work and better public knowledge about the cause of the disease.

Although advances in medical science have had a profound effect on human health, particularly since the beginning of this century, to separate the influence of medical science from other environmental circumstances is not easy. Who can apportion credit for improved expectation of life, as between modern drugs, vaccines and surgical procedures on the one hand, and improved living conditions on the other ? The general approach of medicine has become much wider than it used to be and medical people are now becoming increasingly concerned about the effects of their efforts on the total

W. Harding le Riche is professor and head of the Department of Epidemiology and Biometrics in the University of Toronto. Having grown up in Africa and being professionally concerned with preventive medicine, he is preoccupied with the ecological, cultural and population problems of the developing countries.

welfare of the community. The undoubted contribution that medical science has made to the world-wide population explosion is one clamant example. Another is the medical man's demand for clean water supplies and sewage disposal. In themselves, these are a good thing, yet they lead to a vast wastage of water and, if the sewage-disposal system is inadequate, to large-scale pollution of rivers, lakes and streams. Coping with pollution by sewage is a major problem all over the world today, but particularly in the wealthy countries where the water systems are supposedly the most efficient and highly developed.

Again, medical science is closely engaged in research and action aimed at checking the harm done by industrial air pollution. But industrial processes make our modern world possible, in terms of manufactured goods and jobs for millions of workers. While fighting all types of pollution, we must not kill the industrial goose that lays the golden eggs, nor secure clean air at the price of unemployment, loss of revenue for social services, and consequent hardship, squalor and disease. This is a dilemma from which there is no easy escape for anyone taking a broad view of human wellbeing. In confronting these and other environmental problems, the medical scientist has behind him a long tradition in hygiene and public health. He also encounters, in needlessly sick and dying patients, the sternest of all measurements of environmental mismanagement.

The further that medical science succeeds in tracing the causes of disease, the more intricate are its links with the many components of the human environment. For example, sufficient food of the right kinds is essential for health, yet food poisoning is one of the oldest known causes of disability. It is usually bacterial in origin, but a modern complication is the incorporation in food of many chemicals which may have deleterious effects. Some are used deliberately, such as anti-caking agents, bleaches, maturing agents and many types of preservatives (le Riche 1968). Others occur incidentally; they include many pesticides used in agriculture, safe-guarding food supplies against insects and fungi. If we add to these the many chemicals we use as medicaments there are, at a conservative estimate, at least five thousand such substances ingested by the public. We do not know how many of these may be harmful.

The other essentials of life – air, water, warmth and shelter – are also involved in innumerable diseases, either when they are lacking, so that people are weakened and made more vulnerable to disease, or when these essential supplies themselves are vehicles for infective micro-organisms or man-made pollutants. Medical scientists are therefore interested in every aspect of the domestic environment; also in the social environment and the influence of culture and customs on disease.

In modern societies, people spend much of their lives in offices, mines or factories, and environmental medicine is exercised with questions of safety, bodily stress and noise in these places of work, and especially with ventilation for removal of noxious materials from enclosed spaces. Similar considerations apply to surface transport systems, while passenger aircraft, submarines and spacecraft are examples of micro-environments requiring resourceful attention by the designers to basic medical criteria. The predictions of pollen counts in summer are a daily reminder of another department of our subject – that sensitive reaction of the individual to particular materials in his environment, which we call allergy.

Goitre is endemic in regions where food and water are lacking in iodine; but where natural water is rich in fluoride, dental caries is rare. Such geographic variations in particular diseases are often a pointer to the cause, where this can be traced to a geological excess or deficiency of particular minerals, to a local population of insects transmitting a harmful micro-organism, to contamination from a particular mine or industry, or even to local customs. Certain diseases, most notably heart disease and cancer, principal causes of death in the richer countries of the world, appear to depend upon many factors, some of which are environmental. The same is true for some forms of mental disease, and psychosomatic illness, which involves reciprocal effects of mental and bodily conditions and which is receiving increasing attention as yet another route by which the general environment can affect health.

Despite all these diverse issues, disease caused by micro-organisms is, as it has always been, the central problem of environmental medicine. It is true that, in medically favoured countries, communicable disease is very much in retreat; yet it remains an ever-present threat, especially to our densely populated cities. Among the greater part of the world's population communicable diseases are still rife, causing avoidable death, suffering and debility on a huge scale. Because of air transport and the general increase in travel, a disease originating anywhere can be carried to the other side of the globe within twenty-four hours. For these reasons, and also because communicable disease itself involves a remarkable variety of environmental factors, I shall devote the remainder of the chapter to this aspect of the subject.

Waterborne disease

Much of the developing world is at present finding it difficult to supply the water needed, not only for human consumption, but for sewage disposal from cities, for irrigation in agriculture and for ever-increasing industrial use. To add to the problems, the available water is often contaminated.

Of the many diseases spread by contaminated water, cholera is uppermost in many people's minds at the time of writing, following the serious outbreak in West Bengal and Bangladesh. This region at the mouth of the Ganges is the endemic cradle of the disease, where five great pandemics originated in the nineteenth century, one of them spreading as far as North America. Elsewhere in Asia, there are currently a few cases in Burma, Indonesia, Pakistan and the Philippines, but cholera is also present in North and Central Africa, notably in Ghana, Niger and Nigeria.

Cholera is a serious gastro-intestinal disease characterized by massive diarrhoea, leading to marked fluid loss, dehydration and depletion of potassium and other essential electrolytes. The main characteristic of the disease is its short incubation period of one to three days, rapid onset and death. Cholera organisms (vibrios) are shed into water from the stools or contaminated hands of patients or their attendants. They may survive in well water for three to fifteen days and the El Tor vibrios, a variant strain of the organism, for up to three weeks. They may also survive for up to six weeks in refrigerated foodstuffs such as milk, unsalted butter, meat, fish, oysters and other seafoods. They may also be infective for up to three days on stamps, dry paper, bank notes, leather and metal. An interesting cultural aspect is that, in many places where other ethnic groups are affected by cholera, the Chinese populations seldom get the disease, probably because they use boiled water in making tea and eat freshly cooked food.

For control of cholera, therefore, one needs pure water, filtered and appropriately chlorinated, as well as a high standard of personal cleanliness in the general population, a situation almost impossible of rapid attainment in many of the poor countries of the world.

However, a great effort has to be made, at least to supply pure piped water to as many people as possible.

The benefits of clean drinking water can be illustrated for another notorious waterborne disease, typhoid fever, from the experience of Toronto.

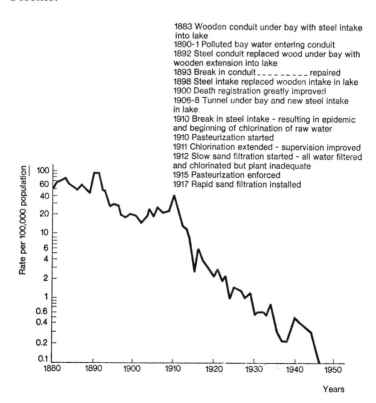

1883 Wooden conduit under bay with steel intake into lake
1890-1 Polluted bay water entering conduit
1892 Steel conduit replaced wood under bay with wooden extension into lake
1893 Break in conduit _ _ _ _ _ _ _ _ _ repaired
1898 Steel intake replaced wooden intake in lake
1900 Death registration greatly improved
1906-8 Tunnel under bay and new steel intake in lake
1910 Break in steel intake - resulting in epidemic and beginning of chlorination of raw water
1910 Pasteurization started
1911 Chlorination extended - supervision improved
1912 Slow sand filtration started - all water filtered and chlorinated but plant inadequate
1915 Pasteurization enforced
1917 Rapid sand filtration installed

Typhoid fever: mortality rates in Toronto, 1880-1972

As the water became progressively cleaner the number of deaths from typhoid declined. This was not the only factor involved, but it was the most important. As with cholera, the typhoid organism passes through human excreta and can infect food. The disease is common in those parts of the world where sewage systems are either non-existent or unsafe.

Historically, cholera and typhoid fever are the best-known diseases spread by foul water, but to these we should also add paratyphoid fever, amoebic and bacterial dysenteries, infectious hepatitis and poliomyelitis. Haemorrhagic jaundice, or Weil's disease, may be acquired by swimming in polluted water; rats are the carriers of the spirochaete responsible for this disease. Eggs of intestinal worms, such as ascaris, may also find their way into water and thus be spread. In Africa particularly, water is contaminated by the larval forms of schistosomes which cause bilharzia, one of the most damaging diseases in that continent.

Bilharzia is an example of how technology has made a disease considerably more menacing by altering the environment. This disease is transmitted from man to snail and back to man, via water, and different parasites are responsible for the urinary bladder form and the intestinal form. The disease is far from controlled in Africa, and there is every prospect that it will increase in prevalence with the growth of irrigation schemes, particularly in Egypt, Nigeria, Ghana and Southern Africa. To break the life cycle through the snail it is necessary to prevent the indigenous populations from urinating and defecating in watercourses. With the rapid increase in populations and the difficulties in improving social services in many of these countries, the educational task faced by governments is formidable indeed.

In those parts of the world where personal cleanliness becomes possible, a great change takes place in the health of the population. For instance it has been observed that diarrhoeal diseases, spread by the faeces-fingers-mouth route, decrease in incidence and prevalence as soon as people start washing their hands, when piped water becomes available. Scabies, caused by a mite, occurs in people who do not wash their bodies and their clothes and it is common during wars, in poverty, and at times of social upheaval. A similar situation holds in the case of the body louse, the carrier of epidemic typhus. Crab lice infest the pubic area. In addition many skin eruptions and fungal infections are common in people who do not wash their bodies.

Animals, parasites and man

In the Western world we assume that clean and uncontaminated food is almost a basic human right. For instance, we expect milk to be produced in such a way that it will not carry disease. However, unpasteurized milk produced on dirty farms, with cows milked by dirty hands or unsterilized milking machines, can carry organisms of typhoid fever, other organisms of the typhoid group, dysentery, tuberculosis, scarlet fever, diphtheria and undulant fever. Ice-cream can easily carry typhoid fever.

When meat is not properly inspected and slaughter of domestic animals goes on unsupervised in unhygienic surroundings, many infections may spread. Chickens are often contaminated with salmonella organisms, resulting in diarrhoea in the consumer, especially if the food is not well cooked. Pork infected with the worm *Trichinella spiralis* may cause trichinosis in humans. From pork we may also acquire the pig tapeworm, from beef the cattle tapeworm, while fish tapeworms are spread from eating raw or poorly cooked fish. The dog tapeworm is found in sheep-raising areas and both sheep and man may be infected by ingesting the eggs of the worm found in the excreta of the dog.

In many primitive societies domestic animals share their owners' quarters and transmit all manner of diseases, including anthrax and brucellosis. Pet-lovers even in prosperous countries are at risk from psittacosis, rabies, ringworm, tetanus and other animal-borne diseases. There are at least thirty diseases, or groups of diseases, which are spread by different types of insects (le Riche and Milner 1971). The common housefly, walking first on human faeces and then on food, can spread any of the diarrhoeal diseases, such as cholera and typhoid, discussed in connection with foul water; the hazards start from the diarrhoeas of infancy and young childhood, which are caused by many different organisms. The virus of trachoma, the blinding eye disease, is also spread by houseflies, while anthrax can be spread by flies carrying spores from infected animals.

Mosquitoes are notoriously the vectors of malaria as well as of yellow fever, dengue and a large group of other virus infections. Plague is spread by the rat flea, while the organism of a disease related to the plague bacillus, tularemia, can be spread by flies, ticks or mosquitoes. The body louse, as mentioned earlier, can transmit typhus. Modern insecticides, particularly DDT, have made a great difference in improving control, particularly of malaria and yellow fever, but their harmful environmental effects are now becoming apparent. We simply cannot, however, ban the use of DDT in Africa or Asia, because it is this relatively cheap material which keeps malaria and other diseases in check. In other words we are in the grip of our own technology, until such time as we can find a better one, perhaps making use of biological control of harmful insects by natural predators or parasites, or by highly selective chemicals such as the insects' own sex hormones.

But we shall be unable, for much longer, to evade a conflict between medical ethics and environmental concerns. As farmer or industrialist, man wantonly destroys other species that share this planet with us, and ruins the Earth's valleys and streams for his selfish ends. Yet man the healer is the most arrogant of all; guided by an honoured ethical code, he shows no mercy at all for any other living creature if a human life is in danger. This is the survival of man at a terrible cost.

The chief means of preventing communicable disease, without wholesale modification of the environment, is the ever-widening range of vaccines made available by medical and veterinary science. Immunization alters the internal 'environment' of the individual so that he can fend off the micro-organisms of disease more reliably. The vaccines in common use cover a score of the most dangerous human diseases, but by no means all, and those that are available are not yet employed routinely in all parts of the world. Even where they are so used, as against smallpox and poliomyelitis in the richer countries, the spectacular diminution of these diseases carries risks of complacency, possibly leading individuals or communities to become lax about the immunization schedules. Unvaccinated people in a country free from a disease are a menace to themselves and others because they will have built up no natural resistance to the disease and a single incoming case could start a major epidemic.

If medicine has an overriding message for the environmental scientist it is this: all dense human populations are permanently at risk from a bewildering variety of microbes and other organisms waiting for an opportunity to strike. The threat cannot be permanently reduced by drugs and vaccines alone, but only by raising to a certain minimum the standards of domestic and public living. To the passer-by, heaps of uncollected refuse may be an affront to his nose and eyes; to the municipal administration, a problem in logistics or labour relations. To the medical scientist, the refuse is nourishment for rats. Rats bearing fleas and bubonic plague once destroyed a great deal of Europe, in the Black Death, and there is no reason why, given the right sort of environmental squalor, they should not do so again.

References and further reading

Robert Arvill, *Man and Environment* (Harmondsworth 1970).
N. C. Brady, *Agriculture and the Quality of our Environment* (Washington, DC 1967).
Fraser Brockington, *World Health* (London 1967).
R. Dubos, *Man Adapting* (New Haven 1965).
W. H. le Riche, 'Food safety as seen by an epidemiologist', *Canadian Medical Association Journal*, 99 (July 1968), 22–7.
W. H. le Riche and Jean Milner, *Epidemiology as Medical Ecology* (London and Boston, Mass. 1971).
G. Wolstenholme and M. O'Connor (eds.), *Health of Mankind* (London 1967).

TOWARDS
SYSTEMS-MINDED
TECHNOLOGY

David Hunter Marks

In the past decade, even as the world's engineers have reached new peaks of technological and managerial skill, typified by the *Apollo* project, there has been serious public questioning of a variety of engineering activities. In some cases this has led to the actual blocking of technically feasible projects. Anyone involved in the provision of large-scale systems such as highways, electric generating stations, dams or supersonic transports knows that the stage upon which such projects must be displayed and evaluated has greatly enlarged and the audience is much more vocal and unpredictable in its preferences. Different segments of society value some objectives over others, and there is now unprecedented conflict between these objectives.

At root, it is a conflict of growth and development, providing jobs, convenience and material objects, as against the desire to preserve the environment and its resources and, by implication, to slow down development. In the United States, for two centuries, the goal has been expansion; only now, with a population of over two hundred million which consumes an astounding proportion of the planet's energy and material wealth, is there sharp awareness of the

David Hunter Marks is associate professor of civil engineering in the Water Resources Division of the Massachusetts Institute of Technology. He previously investigated water-quality management for the United States Public Health Service.

effects on the quality of life of high population densities, unzoned commercialism, ugly tract-housing developments, destruction of natural habitats and environmental pollution.

As a result, the engineer and his clients find themselves beset with paradox. The American public has come to expect unlimited access to electrical energy and no longer regards air-conditioning, for example, as a luxury. But now, despite the ever-increasing demand for power there is also overwhelming opposition to the location of new power plants, which makes it impossible to satisfy the demand. The predictable outcome includes summer 'brown-outs' and power failures, very irate customers and a bewildered but righteous utility management.

The power engineer has for so long been involved only in the details of technology, that he tends to look for more advanced technologies that will keep up with spiralling demand – the 'technological fix'. So far, he has brought little analytical force to bear on questions which are perhaps more to the point. Who is making this demand for electrical power and how might it be modified? How do our present energy policies affect development and hence future demand? What are the environmental limits to energy production?

A classic example of the 'technological fix' was the development of the private car and especially of the highways to carry it. The choice of this major mode of transport was made with virtually no analysis of the role of transportation in society. In the past few decades, United States cities tried to clear up congestion and improve transportation by building more highways. The results have been even more congestion through induced demand, as well as suburban sprawl and decay of city centres, and grave disadvantage, in lack of mobility, for those too young, too old, too poor or too sick to be able to have their own private vehicle. For lack of foresight in this development the engineer shares full responsibility with the politician.

Traditionally, the engineer applies his knowledge of science to deal effectively with practical problems. The test of effectiveness is that returns are supposed to exceed the dollar costs of the project. There are some basic concessions to social values – for example, buildings ought not to collapse except under extreme provocation. But indirect social costs such as water pollution or unemployment enter into design considerations only if they are forced to do so by legislation, company policy or prudence. Within this sheltered atmosphere the engineer has prospered at his well-defined task, with little external control to limit his options.

When the external effects of technological development become great and widespread, the simple criterion of economic efficiency no longer suffices and evaluation becomes more difficult. How is one to measure effects that are not economic in nature – what, for instance, is the dollar value of a human life? Even when effects are reduced to commensurate units different preferences and objectives are operating in society – is a dollar for construction of hospitals the same as a dollar spent for defence or for highways?

From now on, the engineer must be aware of social and environmental consequences of his design and find ways of exposing questions of policy to the widest scrutiny. The whole design process is thereby opened to a host of new issues, many of which are not amenable to the engineer's methods. Institutional problems, for instance, can make the politics of implementation the most decisive element in the design process. In addition, the engineer must be 'systems-minded', concerned not only with his particular problem area but also with the way it fits into the large-scale system in which he works. He has to take a more systematic view of objectives, options, constraints and measures of effectiveness.

The need for new attitudes in engineering is well exemplified by a large-scale system which has many technical and non-technical aspects, and which directly deals with environmental matters. It is the waste-management system, to which I now turn.

The waste-management system

The accompanying diagram, taken from a major study of waste management in the United States (Anderson 1968), demonstrates the principal components of the system. We see three main sectors – waste generation, intermediate treatment and final disposal – and the flows of different types of waste materials between them. This system costs from three to five billion dollars per year in the United States, in direct capital and operating charges, and moves 125 million tons of materials. Thinking just of residential waste production, which is small compared with industrial and commercial waste generation, the average American uses nearly 600 litres of water per day and produces about 450 litres of sewage; he uses about three kilograms of solid materials per day and produces almost as much

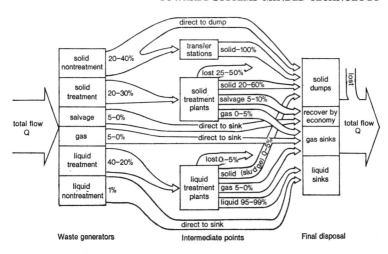

The generalized system of waste management in the United States (after Anderson 1968)

solid waste, most of it packaging materials; and he consumes about one kilogram per day of energy-producing material which contributes about 90 grams per day of particulate matter to the atmosphere (Wolman 1965).

In the various sectors of the waste-management system shown, the options are not always those traditionally considered by the engineer, and there are major interdependencies within the system, and interactions with society. Because the subject is so vast, I shall emphasize one aspect, the disposal of solid wastes.

Waste generation

Waste is generated in three forms: solid, liquid and gaseous. The last goes directly into the atmosphere. Liquid wastes dilute essentially solid household waste so that it can be carried in that very expensive system of underground transport called a sewer; eventually, further great expense may or may not be incurred to get those solids out of solution. Solid wastes are generally collected above ground by a vehicle system which is very labour-intensive.

Fully 80 per cent of solid-waste costs at present arise in collection costs and only 20 per cent in disposal. What are the alternatives? One innovation has been the garbage grinder which makes solid organic wastes such as discarded food transportable by the sewers, thereby shifting the cost from a system with high operating costs to one with high capital costs. Most municipal governments in the United States raise capital funds differently from operating funds and find it easier to get millions of dollars to build something than to find the funds to operate and maintain it. Thus, there is a preference for the capital-intensive, cheap-running alternative.

Other technical possibilities in the waste-generation phase include compactors that reduce the volume of solid waste before transport and, for the future, such things as highly efficient domestic incinerators which would handle all waste at source, or pipelines for moving solid garbage as well as liquids.

Before we resort to ingenious new technologies, there is a neglected range of non-technical options that could greatly reduce the costs of waste management. These centre on reducing the amount of waste generated in the first place, which may in any case represent a tragic misallocation of resources. Much of the solid wastes in the United States are packaging materials, so why not find better packaging? Taxes on certain types of wastes, to reflect their social costs as well as their private costs, could force a shift to less objectionable packaging, while other economic incentives or laws could keep such awkward materials as glass, aluminium and PVC out of the system.

Anticipating the problems of recycling, we see a major difficulty in separating the different materials during treatment. Why should they be combined in the first place? At one time there was a waste-paper industry in the United States which found it profitable to collect old newspapers and other household paper wastes, but this is no longer true. Should not municipal governments be subsidizing such recovery of resources and requiring the separation of waste in the home? With garbage collection in New York City averaging 35 to 45 dollars per ton, it costs almost as much to carry away the used Sunday *New York Times* as residential waste as it does to buy it in the first place.

Intermediate treatment

The engineer's traditional role in waste management is in treatment facilities and in the technological and managerial aspects of bringing the wastes from their sources for treatment. The philosophy has been that of the 'end of the pipe'; that is to say, design has been based on passive acceptance of whatever quantities and compositions of wastes are produced. Moreover, many of the treatment devices are old-fashioned and may just switch wastes from one form to another without accomplishing much in the way of reduction. Solid waste incinerators are major sources of air pollution in most cities that employ them. Meeting current air-pollution standards could increase the cost per ton of waste incineration four to five times.

Computers and systems analysis are now used to deal with some phases of waste handling and treatment in a systematic manner. Mathematical models show how vehicles are best routed or transfer points located for greatest efficiency, while other computer simula-tions investigate different types of work policies. But such measures are palliative only; waste treatment, like the other parts of the system, is dominated by non-technical and institutional problems.

Reclamation and recycling of resources is now attracting much thought and technological investigation. The engineering problems revolve around the uneconomic costs of separating or 'mining' waste and the lack of markets for the separated products. Again, government intervention seems necessary, to subsidize markets or processes for the sake of better use of resources.

Disposal

Society as a whole is most aware of the effects of waste management in the choice of the final sink for the disposal of wastes. Water, land and air have only a finite capacity for assimilating processed or unprocessed waste before there is interference with other uses. Pollution is not an absolute concept – so many gallons of sewage – but needs to be reckoned in terms of alternative uses of the environment it affects. For instance, using a river for waste disposal will at some level interfere with its use for drinking water and at some greater level interfere with wildlife and human recreation. Standards for environmental quality are thus negotiable among various objectives, taking account of physical conditions, public preferences and economic and non-economic factors. As long as development appeared to affect environmental quality only locally, little thought was given to pollution control. Now the public wants pollution to be abated. There is significant economic cost involved here which is passed on to the consumer, reflecting the social costs. Safety measures and air-pollution requirements in United States automobiles are expected to raise vehicle costs by as much as $800 by 1975. Such an impact helps to quantify environmental concerns and the trade-offs between conflicting objectives.

A traditional means of disposing of solid waste has been the open dump, which threatens public health, causes water pollution and impairs the value and use of the surrounding land. Population pressures are making the sites for open dumps virtually unobtainable in large cities. In any case, the burying of solid wastes in carefully managed landfill schemes is plainly preferable, and several studies have shown that landfill is economically feasible even when it involves rail hauls of as much as 300 kilometres (Marks and Liebman 1970). For this purpose, big American cities have actively sought out abandoned coal mines, gravel pits or other rural wastelands, but no such scheme has begun to work; the hang-up is not technical or economic but institutional and social. What small rural area would want to be known as the garbage dump of Boston, regardless of what Boston might be willing, within reason, to pay for that privilege? Only regional management of waste disposal can bring equity between cities and rural areas in distributing the costs, impacts and benefits. Meanwhile technology sits idle, waiting for these non-technical hurdles to be overcome.

System interactions

A strong case can be made for the management of waste as a single system. A systems approach opens a much wider range of technical and non-technical options than have traditionally been considered. This is important because the waste-management system is highly interactive and extreme pressure on one part may well cause problems in other areas. Responsibility for each area is usually fragmented among different units of government and the problems themselves often cross political boundaries. The trade-off between solid waste and water pollution when solids are ground up for carriage by sewer is one example of the interaction between different forms of waste; another is the air pollution involved in the incineration of wastes. Less obvious is the trade-off between organized and non-organized systems for transporting wastes.

Consider the unexpected results of increasing the frequency of waste collection in residential areas. There are public-health and aesthetic benefits if wastes remain in the home a shorter time. In an experiment in Chicago, when collection was changed from once a week to twice a week, the total waste collected increased by 40 to 60 per cent! (Quon *et al.* 1968). The reason for this, I think, is that the organized system captured waste which previously was discarded in an unauthorized manner. In most United States cities as much cost is involved in cleaning the streets and the vacant lots as in refuse collection; storm sewers, too, contain tremendous amounts of large debris which emerge floating in the water after a storm. In the Chicago experiment, more frequent collections shifted these wastes into the organized system – obviously an overall improvement, but one which transferred the cost from street cleaning to waste collection. Municipal budgets can rarely adjust quickly enough to such changes.

Apart from these internal interactions, extraneous effects also impinge strongly on the waste-management system. Changes in steel technology seriously affected scrap steel prices in the United States; this in turn caused the massive abandonment on city streets of junk autos which had previously been recycled. And, as I have mentioned, a lack of markets for other recoverable materials seriously hinders recycling in general.

Institutional arrangements

Adequate management of environmental systems thus depends on the establishment of the proper political institutions which are problem-oriented and can cope with external factors which lead to misallocation of resources. For water resources in the United States, river-basin commissions are already taking over from local political interests and similar agencies are being considered for air pollution and solid wastes. Even with the formation of such institutions, an equitable distribution of economic and social costs has still to be reached and variations in political awareness and leverage between different groups make this process very difficult. The main opponents of change in waste-management systems include the employees and their unions, who make changes in job assignment, in the nature of the work or in measures of productivity difficult if not impossible to implement (Savas 1971).

Where does the engineer fit in? Actually, at many different levels, some traditional, others reflecting a major evolution of his role. New technology is still needed for transporting and treating wastes, for recycling and for more efficient use of resources. Engineers working in other areas which will have environmental impact, for example by generating waste, are conscious of increased public scrutiny. A more definitive role is evolving for the engineer who assists in public decision-making. To his knowledge of technology he adds an understanding of multiple objectives in problem-solving, making him the prime individual to analyse and present the available options for the decision process. The engineer is thus introduced to a new level of policy analysis. In waste management, he is not just an expert on hydraulics or incineration but is interested in regional co-operation, in non-technological controls of waste generation, and in cause-and-effect relationships in the system that may be political as well as technical. Traffic engineers who previously figured out the pavement thickness and the curve radius now evaluate the various transportation options from mass-transit systems to private cars, and appropriate mixes in between. Power engineers not only investigate new technologies that may be environmentally benign but also begin to ponder the limits to the energy that can be produced and used within a given region.

I do not mean to imply that engineers have all the answers; we are just beginning to understand the problem and there is no panacea. The engineer's role is changing in an evolutionary rather than a revolutionary manner, as he becomes more sensitive to the environment and begins to grasp the conflicting objectives of society and their relevance to engineering design.

References and further reading

L.E. Anderson, *A Mathematical Model for the Optimization of a Wastes Management System*, Report 68–1, Sanitary Engineering Research Laboratory, University of California (Berkeley, California 1968).

D.H. Marks and J.C. Liebman, *Mathematical Analysis of Solid Waste Collection*, Johns Hopkins University Department of Geography and Environmental Engineering (Baltimore 1970).

J.C. Quon, M. Tanaha and Charnes, 'Refuse quantities and frequency of service', *Journal of the Sanitary Engineering Division, American Society of Civil Engineers* (1968).

E.S. Savas, 'Municipal monopolies', *Harper's Magazine* (October 1971).

A. Wolman, 'The metabolism of cities', *Scientific American*, 218 (1965), 9.

THE LOGIC OF ENVIRONMENTAL CHOICE

Wilfred Beckerman

The role of the economist in environmental policy is almost universally misunderstood. Non-economists widely suppose, for example, that the economist will inevitably place a lower value on the protection of the environment than would anybody else. More generally, the economist is seen as being concerned with vulgar material output or with the short-run consequences of any choice, and as attaching little importance to higher, more spiritual values or to the longer-run interests of society. But the fact is that the economist, in his professional capacity, has no deliberate pattern of preferences at all, whatever he may think in his private capacity as an ordinary individual. The reason why the economist is often cast as a villain is that he is obliged to point out that choices have to be made between alternative desirable ends. Society cannot, for example, devote unlimited resources to the protection of the environment, however much we might all like to have completely pure rivers, clean air and so on.

Economics being largely the logic of choice, much of it consists of an elaboration of rules of optimum choice, taking account of human objectives and of the constraints upon the extent to which

Wilfred Beckerman has been professor and head of the Department of Political Economy at University College London since 1969. His current research deals with the economics of pollution and he is a member of the Royal Commission on Environmental pollution.

ECONOMICS

these objectives can be realized. The first kind of input from economics into the formulation of environmental policy, then, consists of general principles for considering alternative uses of resources. To input of a second kind, in the quantification of key elements that need to be taken into account in applying these principles, we shall return later.

Constraints on Crusoe

Real-life situations are highly complicated, but the position of the economist can be best explained, perhaps, with the aid of the time-honoured example of Robinson Crusoe on his island. Suppose Crusoe likes listening to the parrots singing and also needs to catch fish for food, but there is a constraint on the key resource available to him, namely his time. Now basic economic theory tells us that his optimum combination of fishing and parrot-music will be that at which the relative satisfactions he obtains from an extra bar of music and from eating an extra fish are proportional to the transformation of music time into time required to catch an extra fish – which is, let us say, one minute. It is obviously pointless for him to sacrifice the music for the fishing so long as the satisfaction he gets from one more minute's music is greater than he would have from eating one more fish. But after a while he starts to feel hungry and his relative preferences change to the point that, even if the technical relationships between music and fish remain the same – one minute equivalent to one fish – it becomes worth his while to go fishing. On the other hand, it might happen that he suddenly noticed a whole shoal of fish come by and realized that in one minute he could catch ten fish. In this case, even if his basic preferences had not changed, it might become opportune for him to switch from music to fish.

In short, Crusoe's optimum choice depends upon two functions: on the one hand, relative preferences for fish and music, which are a matter of subjective taste and can change, and on the other an objective, technical matter of the extent to which time can be transformed from music to fish in the world in which he happens to live. In this simplified example, Crusoe and an economist would have little difficulty in agreeing as to how he should allocate his time optimally. A critic might complain that Crusoe was hedonistic in his parrot-listening or else materialistic about fish, but it would be absurd to criticize Crusoe's economist for giving biased advice, when he had merely told Crusoe how to maximize his satisfactions given Crusoe's particular preferences and tastes.

Of course, the economist's activities cannot be entirely free of value judgements, any more than the scientist's, engineer's or administrator's. Although all he is doing is to advise Crusoe how to maximize his preference function subject to certain constraints, the economist makes a value judgement in deciding that maximizing Crusoe's satisfactions is a desirable objective. This qualification becomes serious when differences of interest within society are introduced into the picture. But otherwise the fact that the economist may advise Crusoe to apportion his time in a particular way implies nothing about the preferences of the economist or of the economics profession in general, as between the spirit and the stomach.

Suppose, though, that a visiting ecologist says that the rate at which Crusoe is catching fish is so great that the fish supply will eventually be depleted, with dire consequences for Crusoe's future food supplies. This is a technical challenge and, in this case, the task is to revise the technical transformation required for deciding the optimum allocation of Crusoe's time. If what the ecologist says is true, then it should certainly be incorporated into the analysis and the economist has no reason to want to exclude it. In economic terms, extra consumption of fish today means not only a sacrifice of music today, but also a sacrifice of fish tomorrow. Indeed, economic theory tells us that policy is optimal only if it takes into account *all* the relevant choices, whether it be between fish and music today, or between fish today and fish tomorrow, not to mention fish today and collecting coconuts today or building a boat today in order to go out tomorrow to a spot where fish will be easier to find.

The implications of fish depletion to which the ecologist has drawn attention nevertheless raise some difficult problems which have long exercised economic theorists. How one should value a unit of consumption tomorrow in order to make it comparable with a unit of consumption today? Again, how one should take account of risk and uncertainty? The ecologist probably cannot be certain exactly how fast fish *will* be depleted at Crusoe's current rate of fishing, while Crusoe himself may have reasonable hopes of being rescued some day from the island before his overfishing can jeopardize his food supplies. Given the uncertainty surrounding the future, how far is it rational for him to go hungry now in order to protect himself against the risk of having trouble finding enough fish in the future?

Economics has been in the forefront of the analysis of the optimum allocation of resources over time. It offers elaborate methods of determining the appropriate rate at which society should discount the future – that is, the extra value to individuals, or society, of consuming something today rather than waiting to consume it (perhaps!) next year. Economics also concerns itself with the search for objective evidence that might throw light on what this rate should be. An equally sophisticated literature deals with the theory of choice under conditions of risk and uncertainty; given our lack of knowledge about many environmental phenomena, this part of economic theory could hardly be more relevant and important.

Here the economist has to act as a guardian of economic reasonableness, seeing to it that people are not unnecessarily impoverished today because of some illusory or hypochondriacal concern about the future. To take but a simple illustration, he may rightly jib at large sacrifices to reduce some particular form of depletion or pollution, when the damage done by it has not been shown to be significant but it is thought that the damage might turn out to be serious. Now, if it cost nothing to insure against such a possibility, then clearly it would be wise to do so. But no rational person should be prepared to spend an unlimited amount today – which means to incur an unlimited sacrifice of current consumption – in order to protect himself against every conceivable risk to which he is exposed in this world. Optimal policy always requires some balance to be struck between the costs of the current sacrifice of consumption and the estimate of the mean probability of the future loss avoided. Human beings are constantly making this sort of trade-off, however unconsciously. Few people would refuse to save anything at all for their old age because they might be run over by a bus tomorrow; or go to the other extreme and save half of their income because of the risk that they might be blinded by lightning.

Public goods and bads

An exceptionally important ingredient of economic analysis, that helps in the formulation of environmental policy, is the theory of public choice and of the allocation of resources between public and private consumption. This is a relatively recent development, but an exciting one and relevant to the shift in social concerns in advanced economies. Economists have long recognized that, unaided, the market mechanism is unlikely to achieve the optimum allocation of expenditure on goods known as 'public goods'.

There are various criteria for defining a public good, but one classic example favoured by writers of textbooks is the lighthouse. In this case, if the lighthouse is provided to serve one ship the light is a free good for any other ships that may happen to be in the vicinity, so that it would be impossible for the lighthouse owner to charge any particular ship with the full cost of providing the lighthouse without allowing all other ships to use it free. Another example is radio broadcasts; once the transmission goes out anybody who has a radio can listen. In both these cases the cost to society of an additional user taking advantage of the facility is zero; one more ship looking at the light does not detract from the amount of light available for the other ships to look at, at least up to the limits of congestion and safety. Hence, market theory requires that the optimal price should be zero, since only at this point do we satisfy the usual marginal optimization condition that the marginal cost to society of further units of the service concerned is equal to the price the user should pay. Under such conditions it is easy to see that the market mechanism is unlikely to provide the optimum amount of the service in question, and some sort of public intervention is needed.

Pollution is a similar case, though not quite as similar as might seem at first sight. For much pollution is a 'public bad' (Winch 1971), in the sense that if one man suffers the smell from a polluted river he does not detract from the amount of the smell that is left for other people to suffer. Conversely, clearing up pollution is often a public good, in the sense that if the public authorities decide to purify the river in order to reduce the health risk to the people who live near it they will, at the same time, and at no extra cost, be reducing the health risk to people for miles around or to anyone likely to visit the river. Economists have always recognized that the evaluation of 'public goods' is a difficult operation; there will be similar difficulties in trying to find any good estimate of the value that people would place on the reduction of 'public bads', as long as they cannot be charged individually for any particular facility or pollution-control measure.

The reason why much pollution prevention should not be treated in exactly the same way as if it were an ordinary public good, requiring the provision of public facilities to clear up pollution, is that pollution can also be reduced by taxing it at source. The same cannot be said of the darkness which creates the need for a lighthouse!

The problem of public goods and bads, which comes close to the heart of environmental policy, leads us to the theory of public choice. In the Crusoe story we were concerned with maximizing only one person's welfare; in practice we have to think of the welfare of millions of people.

Difficulty arises partly because people's preferences may differ, and partly because pollution, or its abatement, affects different individuals and different classes of the community in different ways. Compared with the well-to-do, poorer people are likely to give less weight to environmental considerations even though they usually live and work in a worse environment. The poor may still attach more *relative* importance to having a square meal every day, or a decent roof over their head or a steady job, than to reducing the carbon-monoxide concentration in the atmosphere of the cities (Baumol 1972). A policy to improve the environment may benefit some groups in society while others may actually lose, for example by a reduction in resources that might otherwise have been used for the benefit of the relatively poor.

Such complications have long been the subject of intensive study in economic theory. Besides looking at social welfare in terms of overall national product and economic welfare in general, it is also necessary to consider its distribution amongst the individuals in society. The welfare economics of the late 1930s and the 1940s, in defining an increase in welfare resulting from a policy innovation, took account of the need to be able to compensate the losers. A newer facet is the work on the definition of a social welfare function, given differences in people's preferences. The very basic work by Kenneth Arrow (1963) showed the impossibility of drawing up any unique ranking of social preferences unless one is prepared to make some heroic assumptions about the legitimacy of inter-personal comparisons of utilities.

Yet another aspect of the complexities that arise when one is dealing with several people instead of Robinson Crusoe alone is uncovered in the game-theory approach to economic analysis, initiated by the great mathematician John von Neumann in collaboration with the economist Oscar Morgenstern in their famous *The Theory of Games and Economic Behaviour* (1944), in which the inadequacy of the Robinson Crusoe analogy is explicitly mentioned. Game theory is, in fact, particularly relevant to certain environmental problems, notably those concerned with the management of joint resources, such as rivers, lakes or seas bordered by several countries, or the problems of preventing over-fishing in international waters.

Costs and benefits

The economist is inevitably in the unpopular position of having to remind people that they cannot have everything they would like. As far as environmental policy is concerned, a decision to devote more resources to preserving or improving the environment usually implies that less resources can be devoted to other purposes, such as health, education, housing, clothing, food, leisure, travel and so on. The second main part of the applied economist's job in environmental policy-making is in trying to estimate the relevant 'trade-off' of environmental protection as against other uses of resources – in effect, the relative benefits and costs of improving the environment in any particular instance.

The two sides of this cost-benefit calculation give rise to rather different problems. Reckoning the cost of improving the environment is largely a technical matter, for which engineers and accountants must supply the basic data. How much does it cost, for instance, to use taller chimney-stacks in order to obtain a better dispersion of sulphur dioxide, as compared with removing sulphur from fuels? But even here, the economist has a particular contribution to make. The cost that is relevant to social choice is the true social cost, which is not always the cost as worked out by engineers, policy-makers or firms. Apart from problems such as the correct treatment of capital costs as distinct from operating costs, it is also necessary to aim at cost in terms of the real resources used. Adjustments often have to be made for any element of subsidy or tax included in the normal accounting data. Various other divergences between observed market prices (or costs) and true social costs can be identified only by somebody with a professional understanding of how the economic system actually operates.

But the main contribution of economics to this comparison of costs and benefits usually lies on the side of the calculation of the benefits to be derived from environmental policies. We have to go beyond the benefits that enter into national product (GNP) as conventionally measured. If Crusoe complied with usual practice he would value his fish catch according to the objective data available on the price of fish sold in other markets, perhaps with a mark-up, based on the price of a fish dinner at Pruniers, to allow for his service input into the final cooked meal; but he would not include the parrots' music in the absence of any objective basis for valuing it.

Thus, in evaluating the benefits of environmental protection, economists try to include, in addition to items such as the reductions in costs imposed on the victims of pollution, the gain in welfare to the population from, say, improved recreational possibilities, better health and generally better amenity, irrespective of how far these items have any impact on national product. It is true that this side of a cost-benefit calculation tends to be the weakest; the methodology is in its infancy and the required data are very fragmentary. But in many cases the weakest link in the chain is not the economic input but the scientific input concerning the physical effects of pollution. In most cases, it is not because the economist cannot value some physical effect that the benefits of improved environment are inadequately covered, but because the scientists do not yet know much about the effects of, say, prolonged exposure to low-level concentrations of common air pollutants, of accumulation of pesticides in body tissues, and so on. In this connection everybody is in more or less the same boat; widespread interest in these terribly difficult problems is a relatively recent phenomenon, so that it is not surprising that the data and methods required for solving them are still in an unsatisfactory state.

Another aspect of the calculation that is of special concern to the economist is the cost of alternative policies for environmental protection. A given target for environmental policy is likely to be achieved at minimum costs if this policy provides the same sort of incentives to efficient allocation of resources that is usually provided by the market mechanism. For this reason economists tend to be averse to direct regulation and control as the instrument of policy and to prefer some form of pollution charge or tax. Subject to certain fairly modest assumptions about the way firms behave and the effects, in the longer run, of different rates of profit in different firms, the pollution charge will tend to lead all polluters to take remedial action just to the point where the marginal costs to them of further abatement would exceed the charge. If the charge is set at the optimum level, which economic theory tells us is where the marginal social costs of abatement just equal the marginal social benefits, this implies that polluters will abate up to the socially optimum point. And since the charge should be uniform for all polluters (in relation to the damage done by their pollution) it would not be possible to obtain the same amount of abatement at lower social cost by assigning most of it to those firms for whom the abatement costs are least. Administrators and technologists, however, tend to prefer time-honoured methods of direct control. Many of the reasons that they give for this preference are based on a misconception of the economics of the argument, but they cannot be pursued in this brief chapter.

Perhaps enough has been said to show that, in so far as economics is mainly about choice, it must have a bearing on society's choice between environmental protection and other uses of resources. The fact that this choice may be unpalatable to some in no way detracts from its importance if social problems are to be handled in a responsible manner. Furthermore, many of the salient features of environmental policy, such as the risk and uncertainty aspects, the question of making due allowance for the costs and benefits to future generations, the scope of public goods, the difficulties of evaluating costs and benefits of alternative policies, the allowance for the conflicts of interest within society and the effects on income distribution of any environmental policy, are all matters to which economists are uniquely qualified to contribute.

References and further reading

K. J. Arrow, *Social Choice and Individual Values* (New York 1963).

W. J. Baumol, 'Environmental protection and the distribution of income', *Problems of Environmental Economics*, OECD (Paris 1972).

Wilfred Beckerman, 'Economists, scientists and environmental catastrophe', *Oxford Economic Papers* (November 1972).

Wilfred Beckerman, 'Environment, "needs" and real income comparisons', *Review of Income and Wealth* (March 1973).

Wilfred Beckerman, 'Environmental policy and the challenge to economic theory', *Political Economy of Environment: Problems of Method*, ed. Ignacy Sachs (Paris and The Hague 1972).

Council of Economic Advisers, 'Economic growth and the efficient use of resources', *1971 Annual Report USA*, chapter 4 (Washington, DC 1971).

Allen Kneese, *Water Pollution: Economic Aspects and Research Needs* (Washington, DC, 4th printing 1970).

John von Neumann and Oscar Morgenstern, *The Theory of Games and Economic Behaviour* (Princeton 1944).

Royal Commission on Environmental Pollution, *First Report*, chapter 2, Her Majesty's Stationery Office, Cmnd. 4585 (London 1971).

The Swedish Journal of Economics, Special Issue on Environmental Economics (March 1971).

D.M.Winch, *Analytical Welfare Economics* (Harmondsworth 1971), p. 123.

FROM MAPS
TO MODELS
A. G. Wilson

The nature of geography has changed so much, and is continuing to change at so great a rate, that most people with memories only of school geography would not recognize it. Geography nowadays is seen as having two major themes: the methods of analysing the distribution in space of objects and activities, and an integrating approach to certain systems of interest. It therefore seems appropriate to begin with some account of geography as a modern discipline. The main changes have been generated by advances in theoretical geography, an outline of which forms the second part of my argument. Thirdly, the new approaches are illustrated by the example of people in an urban environment where it turns out that current responses to environmental problems reflect only partial analyses, although methods are being developed for more comprehensive treatment. In conclusion, I shall return to the broader question of what can be expected from geographical theory as a whole, as a contribution to environmental science.

The new geography

It is customary to begin discussions of this kind with Hartshorne's definition: 'Geography is concerned to provide accurate, orderly,

A. G. Wilson is professor of urban and regional geography in the University of Leeds. He held previous appointments as assistant director of the Centre for Environmental Studies in London, and mathematical advisor at the Ministry of Transport. He is the author of *Entropy in Urban and Regional Modelling* and *Papers in Urban and Regional Analysis*.

and rational description and interpretation of the variable character of the Earth's surface'. This clearly reflects geography's traditional role of mapping and describing the Earth's features, whether physical or human. The geographer would only rarely, of course, study the Earth's surface as a whole. More typically, his concern would be some region – a nation, part of a nation, a city, a forest, a drainage basin, or whatever – and its relationship with other regions. Each region, though, is made up of a number of components which interact with each other, and with components of other regions. Any such collection of components can be usefully called a 'system of interest', or more briefly, a 'system'. This is simply a name, which sounds impressive but does not immediately mean much more than a term like 'thing'. To show that the system is a useful concept is one of my tasks.

The geographer, in modern parlance, is thus concerned with systems of interest, which will usually be regions of the Earth's surface; he will be particularly concerned with the present and changing arrangement in space of the components of his systems. The nature of geography's contribution to environmental science in relation to other disciplines is now plain. Clearly, many other disciplines will share an interest in some of the geographer's systems. For example, geologists have an interest in land forms, biologists in forests, economists and sociologists in cities. A discipline is a group of people using a certain body of concepts in relation to particular systems of interest. The geographer shares his systems of interest, and lends and borrows some of his concepts, and hence there are considerable areas of overlap with other disciplines, particularly with respect to the broader coalition of environmental scientists. The overlap is perhaps greatest in the case of new specialist coalitions with such names as 'regional science' or 'urban studies'. The core of geography, though, remains as a special contribution in its own right: spatial analysis and a concern for an integrated understanding of regional systems.

The transition from geography as 'orderly description' to geography as 'understanding systems of interest' may seem a gentle one. It represents, however, a transition from geography as arts subject to geography as science subject, from a concern with something more akin to historical explanation, to the realm of scientific explanation. Most important of all, this kind of scientific explanation provides the basis of a *predictive* capability for systems of obvious relevance to environmental science, and hence a potential problem-solving capability. The overriding concern of the environmental scientist is problem-solving action; the geographer, in appropriate systems of interest, can thus provide much of the analytical basis for action. This is a big claim, so let us now turn to the development which makes it possible.

Theory in geography

The development of theory in geography has a history that goes back to significant contributions in both the physical and human aspects in the nineteenth century. The major advances, however, have come in the past twenty years, with such acceleration that many facets of the subject have changed fundamentally in the past decade. Physical geography, of course, has long benefited from its scientific tradition, shared with geology and meteorology. Human geography, though, has had to make a rapid and sometimes painful transition from an arts to a science base. Hindsight suggests that a surge of theoretical advance should have been possible long before it actually took place. The triggering event was probably the advent of the computer in the 1950s. It enabled geographers to handle large quantities of numerical data and provided the basis for an integrated scientific advance in such fields as urban and regional geography, on both empirical and theoretical fronts.

As in most of the sciences, geography uses two types of theory-building activity – inductive and deductive. The inductive scientist searches for generality by inference, more or less directly from data, usually employing techniques of statistical analysis. The deductive scientist postulates general theories, or devises mathematical models, and tests their particular application to real situations where data is available; as necessary, he modifies his theory as a result of this experience. The inductive method is reliable in the sense that it usually produces some useful results, and safe in that it does not necessarily demand great powers of invention in order to be realistic. The second method demands greater creative abilities, and is thus more risky, but when successful is more likely to produce powerful results. The two methods, and mixtures of them, work hand in hand.

After its quantitative revolution, geography at first followed other social sciences with similar problems, in putting too much emphasis on inductive statistical methods. Only in the past five to ten years has work with models become commonplace. Parallel advances in this technique in other sciences have helped us; indeed, the similarity of the model-building problems in many fields has led to the emergence of a general systems theory. Already geographers have achieved sufficient progress in model-building to see the exciting possibility of even more effective developments in the future.

Examples from the urban environment

Many, perhaps most, human environmental problems are related to the fact that more and more people live, work and play in cities. Congestion, pollution, dereliction, depletion of resources and many other problems arise from the activities of individuals and organizations in urban areas. This in itself is an important point in the debate on the environment: preoccupation with the 'natural' environment often leads to neglect of the overall structure of the man-environment system. Oversimplified 'return-to-nature' proposals cannot solve the problems of the cities and hence can scarcely touch those of the environment in general.

In keeping with my previous argument, the aim of geographical theory in this context is to make available explanatory and predictive models showing why cities are as they are, how they are changing and how they can be changed for the future. A brief discussion of two major urban problems will illustrate what is needed.

Firstly, consider traffic congestion, and associated questions of air pollution. The pattern of urban traffic is determined by the distribution of people, economic activity, jobs, and such things as recreational opportunities in the city. Traffic flows are largely interactions between home and job and car congestion occurs because many people choose to travel by car instead of by public transport.

Models are available which help in the analysis of this problem, and in testing proposed solutions. Given a distribution of people and activities, quite accurate predictions can be made of flows between each pair of points in the city, of the means of transport used and the routes followed. Models can also be built to predict the associated patterns of air pollution. Possible ways of reducing air pollution due to car congestion include:

(i) rearranging the distribution of population and their activities;

(ii) improving public transport facilities; (iii) improving or not improving (for it could be either!) highways or parking facilities for cars; (iv) replacing petrol engines by electric-powered engines. Some mixture of these and other measures may be best – but which? Any combination of policies can be fed into the computer model of traffic flow and the impact of these changes assessed. The package of changes can be chosen which most nearly achieves the desired goals.

Two important caveats are in order at this point. It will not be easy to discover and specify the 'desired goals'; different sectors of the community will almost always have goals which are in conflict. Thus, to say that researchers using models will help solve environmental problems is not to say that they will achieve this on their own. At best, they will inform, and perhaps even transform, the usual political processes. Secondly, the models which are used must accurately reflect and anticipate human behaviour. Many alleged solutions to environmental problems – for example, 'all will be well if x per cent travel by public transport' – are not feasible if people will not behave that way. A good model should be able to check the behavioural feasibility of proposed solutions to problems.

For the second example, consider the depletion of natural resources. Problems arise because scarce natural resources are consumed at short-run prices which do not adequately reflect the long-run shortages, although the low short-run prices may reflect confidence that, in the long run, technological substitutes for the disappearing resources will be found.

Natural resources form the basic inputs to many sectors of the economy. The spatial distribution of all kinds of economic activity tends to follow the locations of available resources and the places where they are used. Models can be built, both of resource utilization within the overall economic structure – these are typically so-called input-output models at national or regional scales – and of the spatial distribution of economic activity. Possible solutions to the depletion of resources lie in new policies of pricing or regulation. Again, any changes can be fed into the economic models and their impact assessed, both in terms of the economy as a whole and in terms of the spatial distribution of activity. A purely economic approach, neglecting geographical factors, would be liable to cause hardship and unexpected environmental effects in many places.

The models referred to are available but are not yet in common use. Even if they were, the problems would almost certainly still be dealt with separately by different agencies of government, despite the evidence that they are strongly interdependent. If a city adjusted its transport system to solve the problems of car congestion and pollution problems, the resulting changes would affect the travel of people to work and links within industry, and hence the spatial distribution of economic activity and, perhaps, the pattern of resource use. If a major technological change were made, in the nature of car engines for example, then the resource-using characteristics of the economy would change. Conversely, if policy adjustments were made in connection with resource usage, this would affect the economy, the spatial distribution of economic activity, and hence the traffic pattern.

This mutual interdependence of apparently dissimilar environmental problems is part of what would be shown up by more thorough analysis: namely that, in urban regions, almost everything depends on almost everything else. Although, of course, detailed specialist work will always be needed, sets of simple solutions to environmental problems, considered in isolation from one another, will rarely be adequate; most current suggestions seem to be at this stage. A general model has to be developed that represents all the interdependencies, and this establishes an analytical requirement which is basic to geography's contribution to environmental science. The components of the necessary general model for the problems of people in an urban environment are summarized in the diagram opposite. Various versions have been built of most of the sub-models shown, some being much more advanced than others; there is relatively little experience, as yet, of the construction of any such general model (Wilson 1972b).

Problems for the future

The ongoing scientific problems can be discussed in relation to a chain of activities concerned with research, with development and with the application of results and tools already available. Typically in such chains, more people are involved in development than research, and more in application than development. Nevertheless, research feeds development and development feeds application and hence problem-solving, so care must be taken with research and development sectors, and particularly with research. Although research problems are difficult to define, governments have somehow to find ways of sponsoring research activity, the benefits of which may be most unclear at the outset, and the risk of failure quite high. Even in long-established branches of science this issue is still under continuous review (UK Government 1972).

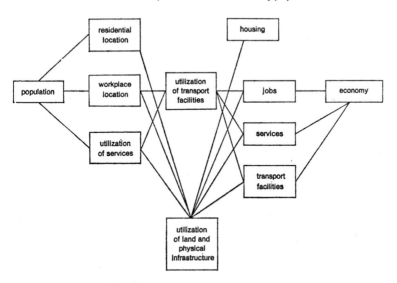

Components of a general model for the problems of people in an urban environment

Some research in geographical systems-modelling is carried out by planning agencies – perhaps because they are forced into it – but, in the main, research is for the time being concentrated in university departments, with some help from relatively small establishments such as the Centre for Environmental Studies in London, and the Urban Institute in Washington, DC. Substantial progress has come from this organizational framework in the past and will continue in the future. For some purposes, however, research will need 'big technology' in support, analogous to the physicists' accelerators but in this case 'soft' technology in the form of environmental information systems replete with numerical data (Wilson 1972c). At the research frontiers at present, and perhaps for the first time for many years, theory-building seems to have run ahead of the information capacity needed for essential empirical development. Governments will have to consider providing large information-system laboratories for the environment to help solve this problem; such laboratories would also provide a focus for interdisciplinary work on partial models and, ultimately, a base for effective general model-building. They would give a tremendous stimulation to innovative work in universities and elsewhere.

Attention also needs to be given to development and application. At present, the best available techniques for environmental planning are not usually used by the agencies concerned, partly because of inadequate development effort, and partly because of the inherent conservatism of the planning professions. Here again, only effective government action can put matters right.

Although I have emphasized, by way of illustration, model-building in urban and regional geography, the same concepts are applicable, *mutatis mutandis*, to other fields of geography in their contribution to environmental science. In the broadest possible terms, the task of the geographer will be to produce a general *world* model – not a single gargantuan model, but a set of linked models each built for a different spatial scale. The amount of detail would vary according to scale; the world model itself would embrace total resources and population, but would be based upon continental, national, regional and urban models. Models may also be useful at more 'micro-spatial' scales, for example for a community within a city. Models in different places in the spatial hierarchy are inter-dependent and the community model, for example, would have to function within an overall 'environment' provided by an urban model.

We might also seek integration for models of overlapping systems of interest. For example, a physical geographer's model of a drainage basin may be integrated with a resource geographer's model of a water supply system and this, in turn, with a human geographer's model predicting water demand. On the whole, this kind of progress has yet to be made.

Geography, then, has almost transformed itself into a fully scientific discipline. Its systems of interest and its concern with spatial analysis are such that, in collaboration with other disciplines, it has a major role to play in the development of diagnostic, pre-dictive and planning capabilities for environmental science. A great deal of effort, and government assistance, are needed if the vision of future advance is to become reality before crucial decisions have to be taken which are analytically ill-informed.

This discussion, elementary though it has been, also illustrates the massive analytical task facing environmental science in general.

Recent adventures in the field of general systems modelling in relation to the environment, and in particular the global reports from the Massachusetts Institute of Technology for the Club of Rome (Meadows *et al.* 1972) serve to illustrate this point in another way. Considerable effort and ingenuity have gone into the model-building task, but the final result does not stand up to scrutiny from the scientific community. Reactions to the MIT work serve as a useful warning that much more scientific work in the field is needed before models can play their proper part in guiding man's management of his global environment.

References and further reading

D. H. Meadows, D. L. Meadows, J. Randers, W. W. Behrens, *The Limits to Growth* (New York and London 1972).

UK Government, *A Framework for Government Research and Development*, Cmnd. 4814 (London 1972).

A. G. Wilson, 'Theoretical geography – some speculations', *Transactions*, Institute of British Geographers (1972a).

A. G. Wilson, *Papers in Urban and Regional Analysis* (London 1972b).

A. G. Wilson, 'Environmental disasters need more than the Blueprint for Survival', *Times Higher Education Supplement* (11 February 1972c).

PART II
COPING WITH
COMPLEXITY

Editor's introduction
to Part II

A common style of working begins to emerge for environmental science, despite its diversity of inputs. Men in muddy boots sample the components of the environment and festoon our planet with instruments. The same men or others sit at computer keyboards trying to elaborate conceptual models of environmental systems. The model is equivalent, in complex systems, to the hypotheses and laws of the 'simplistic' physical sciences.

Although there is emphasis on methodology in this second part of the book, the material is not rigidly all of that kind. Interwoven are chapters that fill out a tolerably coherent picture of some of the environmental systems with which we have to deal, and continue from Part I the theme of expertise, where this has very general applicability. For a start, two chapters look at natural systems through the eyes of physical science, to see the great global processes of physical chemistry (Chapter 14) that set the stage for life and also the operation of fundamental laws of thermodynamics (Chapter 15) that govern all energy transactions, from the molecules that capture sunlight to the dependence of the predator on its prey.

Although great numbers of processes are at work in an environmental system, a major tactic (Chapter 16) is to try to identify one or two limiting factors operating in a given situation. Finding out these key factors can be greatly aided by experiment and, as Chapter 17 illustrates, experiments with 'life-size' environmental systems are by no means as unlikely as one might think. Experiments also figure, together with unwitting perturbations by man, in Chapter 18, as scientists elucidate the ways in which interactions of species give eventual stability to the populations of all. Chapter 19 pursues a practical aspect of populations, namely the sustainable yields that man as a predator can derive from renewable resources – in this case, from whales.

Comprehensive model-making for particular ecosystems is well illustrated by the work on North American grassland described in Chapter 20. This leads us into computer modelling of the regional and global environmental systems of man (Chapter 21) which represent a summit of present methodological ambitions in environmental science.

Then we return to three major environmental issues of our time: for developing countries, the ecological aftermath of the Green Revolution (Chapter 22); for the rich countries, the questions of pollution associated with energy supply (Chapter 23); for the whole world, the overall assessment of man's resources in terms of those which are inherently local and those that are transportable and must be shared (Chapter 24). Emergency situations, typified by the wreck of *Torrey Canyon*, as discussed in Chapter 25, demand a special kind of response from research scientists.

Chapter 26 describes present efforts to institutionalize teaching and associated research in environmental science, with special reference to experience in the University of California. Finally in Chapter 27 the editor, looking to the future, exhibits his prejudices.

Nigel Calder

EQUATIONS OF SURVIVAL

Ferren MacIntyre

Mother Nature: the personification of the environment that nurtured mankind is not quite as naive as is sometimes thought. In important respects, nature is like a living organism, growing, evolving, but nearly always in balance. In the environment, as in an organism, critical conditions are kept under various forms of control which preserve an essential constancy. The comparison is with 'homeostasis', the maintenance at constant values of the pressure, temperature and composition of the blood, or other inward circumstances on which the life of an organism depends. The best evidence for long-term uniformity in conditions on Earth is the existence of life itself, for we organisms are fragile and can cope with only very small changes in our environment.

Some species, 'living fossils', have survived from earlier epochs because they are 'so well adapted to a particular *continuously available environment* that almost any mutation . . . must be disadvantageous' (Simpson 1944, my italics). Thus the survival of the horseshoe crab bespeaks a seashore essentially unchanged for 200 million years. Dome-like sedimentary structures (stromatolites) built by shallow-water algae extend back for some 2,800 million years, while microfossils resembling algae are now known from 3,200 million years ago. Their discovery suggests that at least some

Ferren MacIntyre is a physical chemist, and research associate at the Marine Science Institute of the University of California, Santa Barbara. His research centres on the sea surface microlayer and its role in the exchange of material between the ocean and atmosphere.

environmental conditions in the ocean have remained approximately constant for the greater part of the Earth's history, although we know that oxygen was formerly much scarcer. Conversely, if any of a dozen variations in the Earth's chemistry at any time in these past 300 million years or so had impaired the tenuous layer of ozone high in the atmosphere, most life on land would have been destroyed by ultra-violet radiation from the Sun. We would not be here. In short, we are utterly dependent on a durable but intricate system of chemical regulators, of which we have been largely unaware until this century.

Mechanisms of regulation

Conceptually, the simplest long-term regulator of the global environment is *chemical equilibrium*. Consider two inter-connected examples (pH and carbon dioxide) given under this heading in the table of global regulators (opposite). If the ocean were alkaline, it would dissolve the carbon dioxide from the atmosphere and plants would be unable to grow. If it were acidic it would attack the limestone on the ocean floor; carbon dioxide would bubble out of the ocean and animals would suffocate. Yet these menacing chemical reactions are a clue to how nature maintains neutrality (pH = 7) in ocean water, because they are reversible. Exchanges of carbon dioxide between air and water, and between water and sediments, proceed in either direction to just the extent that is necessary to correct for any temporary deviations from the norm. Over longer time-scales, this regulator is reinforced by a more powerful one: the ability of clay on the ocean bottom to change its chemical structure and either absorb or release hydrogen ions, thus correcting for acidity or alkalinity.

Chemical equilibrium is the most reliable kind of global regulator. It is characterized by predictable values and an ability to return to normal after a displacement in any direction. It would be expensive and rash to try to test global systems directly for equilibrium behaviour, but when we examine specific reactions in the geological record and in the laboratory, and extrapolate, the results suggest that the systems indicated are in geochemical equilibrium on a 10,000-year time-scale. Appreciable disequilibrium is possible locally or over short periods.

Chemical systems which maintain their composition by balancing

The principal global regulators arranged according to type, as discussed in the text

Under 'time-scales' the 'duration' is the order of magnitude of the period for which conditions remain roughly constant; the 'response' is the time necessary to correct small, temporary deviations

| | Time-scales | |
	duration	response
Chemical equilibrium		
oceanic pH (acidity v. alkalinity)	3,000 million years	10,000 years
concentration of carbon dioxide in air and oceans	3,000 million years	10,000 years
concentration of salts in oceans	3,000 million years	10,000 years
concentration of heavy metals in oceans	3,000 million years	10,000 years
Steady-state		
ultra-violet reaching Earth's surface	500 million years	seconds to years
excess of oxygen over carbon dioxide	500 million years	months
thickness of soil	10 million years	1 million years
Almost steady-state		
concentration of oxygen in air	100 million years	uncertain
area of oceans	100 million years	uncertain
depth of oceans	100 million years	uncertain
mean temperature of planet	10,000 years	25 years
diversity of living species	10 million years	1 million years
Stability from planetary dynamics		
ocean circulation	—	
climatic patterns	—	
No known global regulators		
frequency of volcanic eruptions	—	
formation of coal, oil and natural gas	—	
occurrence of radioactivity	—	

a large input with an equally large output are described as being in *steady-state* (sometimes 'kinetic equilibrium'). An analogy is with two streams which maintain a lake at nearly constant level despite large variations in the input flow; if the level of the lake tends to rise, more water flows out of it.

145

Both in chemical equilibrium and in steady-state systems, control is achieved by *negative feedback* – a reaction to change that exactly compensates for and corrects the change, as in the analogy of the lake and its streams. The difference is that chemical equilibrium is entirely predictable (in principle) and depends on fundamental laws governing the behaviour of materials, while steady-state is more arbitrary and its level is set by the circumstances of the system. A case in point is the amount of ultra-violet radiation reaching the Earth's surface from the Sun, which is governed by the ozone layer already mentioned. If the ozone diminished, more ultra-violet would come through; then it would act on the oxygen of the air to make more ozone and restore the status quo. But the point of balance depends critically on the chemical composition and physical conditions of the atmosphere.

Other global systems, which today seem to be in steady-state, have actually changed slowly and progressively, and I list them as 'almost steady-state'. The diversity of living species, for example, has increased steadily during evolution, despite the alternation of slow increase during times of plenty and quick decline in times of stress.

Intermediate between the well-regulated and unregulated natural systems are those to which complex interactions at the level of planetary dynamics give a measure of stability. The most important are the circulation of ocean water and the pattern of global climate, which in turn are closely interdependent. The currents change as ocean basins grow or shrink and the climate can make excursions into ice ages; yet, over long periods, conditions will remain almost constant.

The ocean as the primary regulator

The ocean is involved in half of the entries in the table on page 145 and it can be thought of as the planet's principal regulator. Ocean water is so uniform that the major oceanic provinces are distinguished by temperature differences of a degree, and salinity differences of 0.1 per cent. This uniformity persists despite a continual input, from rivers, of water of very different chemical composition, and despite a gradual thickening and extension of continents and the concomitant deepening of the oceans. The reason is concealed in the following set of recipes for making land, sea and air.

   i. deep lying rocks + vulcanism → surface rock + *water + gases*

  ii. surface rock + rain water → stream water + detritus

 iii. stream water + detritus → seawater + clay

 iv. clay + heat + pressure → *continental rocks*

Each step results in a nearly uniform product. Rain water picks up carbon dioxide from air and soil and becomes a weak acid (carbonic acid) capable of slowly dissolving alkali metals out of rocks and leaving behind aluminosilicate detritus such as clays. But rivers also carry suspended detritus out to sea and there the aluminosilicate recombines with other metal ions in the water to form oceanic clay. In the water are left the major ions characteristic of seawater, especially sodium and chloride. Eventually the moving ocean floor carries the sediments under the edge of a continent where they are cooked into shales, marbles, granites, and other typical continental rocks. When these emerge, the rain falls on them and the cycle starts again.

These processes can be filled out with a wealth of detail to correspond to, say, a specific volcanic eruption or a particular oceanic clay, but averaged over the globe they revert to the skeleton reactions just described. Because the ocean receives inputs from all continents and mixes them instantaneously (on a geological time-scale) and because its outputs to the sediments are the products of chemical equilibria, ocean composition should be uniform in time and space. This seems to be true even when great changes occur. Nowadays most carbonate is extracted in combination with calcium, in coral reefs and diatom tests, and there has been a suggestion of a relative reduction in oceanic calcium about 440 million years ago, when corals first flourished; but the evidence is not compelling. Nor is there any support for the Victorian idea that the sea gradually became more saline throughout the Earth's history. Ocean water is fully recycled through rain and rivers every four thousand years, giving ample opportunity for leaching additional salt from the continents; yet attempts to find a geological record of a progressive increase in salinity of ocean water have not been successful. One is led back to the fundamental recipes set out above and the chief interest in oceanographic chemistry and palaeochemistry lies in the small and momentary departures from uniformity.

The antiquity and precision of this system, safeguarding the conditions in which life evolved, are somewhat awesome. It is correspondingly thought-provoking to find that man is capable of interfering with these skeleton reactions. In New England, Scandinavia and no doubt in other places downwind from industrial areas, the principal acid in rain water is no longer carbonic acid but sulphuric acid. It is too soon to evaluate the effects of this change on the fundamental process of chemical weathering.

Disrupting the steady-state

Man has been outpacing some regulators, at least locally, ever since farmers began eroding soil that takes many human lifetimes to renew itself (see Thomas 1956). Perhaps the most sobering of man's interferences is with species diversity. The maximum natural extinction rate, during the abrupt decline of the dinosaurs, was one species every thousand years. Today's rate is one species every nine months. The following table summarizes actual and incipient human influences.

Human impacts on global regulation

---

1 Demonstrable environmental damage affecting, at least locally:
   concentrations of heavy metals in oceans
   thickness of soil
   diversity of living species
   populations of living species
2 Demonstrable changes of unknown import affecting:
   concentration of carbon dioxide in air
3 Possible damage in the near future affecting:
   ultra-violet reaching the Earth's surface
   mean temperature of planet
   ocean circulation
   climatic patterns
4 Novel causes of damage for which no natural regulators exist:
   man-made radioactivity
   non-biodegradable organic compounds (e.g. DDT, plastics)

For an example of how easily technological man can now interfere with global regulation, let us consider once again the ultra-violet radiation reaching the surface of the Earth. Had its intensity doubled at any time since life occupied the land, many sensitive species would have been wiped out. Had it decreased much, no animal like ourselves dependent upon ultra-violet to synthesize vitamin D would have survived. From this we may conclude that the control, the abundance of ozone and other ultra-violet-absorbing gases in the upper atmosphere, has remained in steady-state for this period.

But this system is delicately poised and it could be unbalanced by relatively small man-made inputs, because the upper atmosphere is rarefied. The effects of the exhaust gases of a fleet of high-flying supersonic airliners upon stratospheric ozone were considered by the Massachusetts Institute of Technology's study of *Man's Impact on the Global Environment* (SCEP 1970). One calculation was that the added water vapour would reduce ozone and increase ultra-violet by an insignificant 4 per cent. The study-group also concluded – prematurely – that the nitrogen oxides produced in the jet engines 'would be much less significant than the added water vapor and may be neglected'. A later review of the problem (Johnston 1971) found it to be somewhat more complicated than it had appeared, with thirty-one interrelated chemical reactions at work. Each reaction proceeds at its own rate, and the times needed for it to reach steady-state throughout the atmosphere range from a few seconds to several years. Nine of the reactions absorb radiation and therefore influence the stratospheric temperature, which in turn affects all of the reaction rates and also the stratospheric circulation, introducing complications which have not been examined. Two of the reactions of nitrogen oxides decompose ozone catalytically, which means that a single molecule of nitrogen oxide can go on destroying ozone indefinitely. By increasing nitrogen oxides far above the background level, the supersonic fleet could reduce the ozone shield and increase ultra-violet not by a tolerable 4 per cent but by a scorching 50 per cent.

The oxygen supply

Fears about human ineptitude are not always so well founded. In recent years there has been an unwarranted worry about depleting the world's oxygen supply by consuming fossil fuel. A brief examination of the history of oxygen removes this particular fear.

The primitive atmosphere contained only traces of oxygen, produced when solar radiation dissociated water vapour:

i. water + energy → hydrogen + oxygen

For most of the Earth's history, oxygen remained at a low, steady-state concentration because of processes that consumed it as fast as it was made:

    ii. oxygen + water + nitrogen + energy → nitrates
    iii. oxygen + hydrogen sulphide (from volcanoes) → sulphates
    iv. oxygen + ferrous oxide (in minerals) → ferric oxide

Atmospheric oxygen began to increase (in 'almost-steady-state') only after the evolution of plants, which grow by photosynthesis:

    v. carbon dioxide + water + energy → organic matter + oxygen

Respiration, which supports plant life in the dark (and quite incidentally supports animals and other dependent species), reverses the previous reaction:

    vr. organic matter + oxygen → carbon dioxide + water + energy.

These last two processes are almost exactly balanced and in steady-state. Virtually every plant cell is eventually eaten, decomposed or burned, and when that happens it consumes exactly as much oxygen as it produced when it grew, and gives back exactly as much carbon dioxide as it extracted from the atmosphere.

If this were the only possible fate of dead plants, the world's oxygen supply would never have reached its present proportions, but a small percentage of plant material falls into oxygen-free water, such as a bog or a closed ocean basin. Some decomposition occurs:

    vi. organic matter → fossil fuel + carbon dioxide + water

This process, while sparing oxygen and allowing it to accumulate in the atmosphere, leads to coal, petroleum or natural gas – and also to tar sands, oil shale, and vast quantities of shale and slate tinted grey by dispersed carbon. Such buried carbon as we can extract economically, we will burn, thereby completing the decomposition that was interrupted when it was buried:

    vii. fossil fuel + oxygen → carbon dioxide + water + energy

It is the dispersed carbon alone that saves us; we can get at such a small quantity of the total even by consuming all known reserves of fossil fuel that, although we will briefly triple the atmospheric carbon dioxide, we will reduce oxygen only insignificantly, from 21 per cent of the atmosphere to about 20.8 per cent.

In a similar way we can deal with other fears of this kind; for example: 'If synthetic foods become widespread, there can be a geologically rather rapid depletion of atmospheric oxygen'. But synthetic food consumable by humans or other animals must be based on reduced carbon like that made by plants. There is no way of making organic matter from carbon dioxide or limestone without *releasing* oxygen. The only other substantial source of carbon is fossil fuel (as in current experiments with yeasts grown on petroleum) but any such use merely competes with combustion as a consumer of fossil-fuel resources, and the total oxygen consumable remains strictly limited.

As significant for the oxygen steady-state as photosynthesis and respiration are the reactions ii and iii. Reaction ii, nowadays mediated by nitrogen-fixing bacteria, as well as by upper-atmosphere processes, is capable of converting the entire supply of atmospheric oxygen into nitrate in less than ten million years, while a bacterial variant of reaction iii reportedly converts half of the world's supply of atmospheric oxygen into sulphates in a mere fifty thousand years. Everything therefore depends on the reverse reactions:

ii r. nitrate + organic matter → nitrogen + water + carbon dioxide

iii r. sulphate + organic matter → hydrogen sulphide + water + carbon dioxide

These reactions are carried on by other bacteria in soil and in oceanic mud. We can be thankful that the bacteria are not as sensitive to DDT as was the peregrine falcon!

The global mean temperature

Large systems frequently look simple at first, when we do not understand them at all, and simple again much later when we know which are the important aspects and which can be ignored. In between they look very messy indeed. A prime example is the system which controls the planetary mean temperature, T. This system has received much attention lately because of its sensitivity to man's inputs and the suspicion that changes in T of one degree centigrade or less can result in major shifts in climate and global vegetation. As the following diagram may convey, there is nothing simple about it, at this stage of investigation.

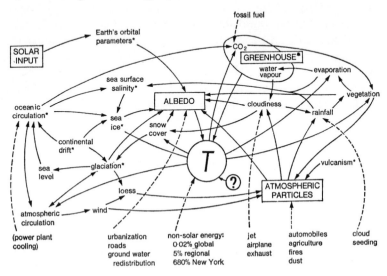

The global regulatory system of the planetary mean temperature T, as at present poorly understood

Some of man's more important contributions are indicated by broken lines. The central question mark, and other features, are discussed in the text

We know from first principles that the Earth's heat loss must exactly balance the input of heat from the Sun plus the relatively insignificant heat coming from other sources. This tells us nothing about the resulting surface temperature. About 30 per cent of the input energy is returned directly to space; this is the *albedo*, the average reflectivity of the planet, and its effect alone would make T approximately $-20°C$. Obviously anything which affects the albedo – notably changes in ice and cloud cover – will affect T also.

The remainder of the heat coming from the Sun is absorbed and must be disposed of by reradiation from the relatively cool Earth – which means by infra-red radiation. Gases in the atmosphere impede the process by reabsorbing the radiation except in particular infra-red-transmitting 'windows'. By blocking long-wave infra-red, these gases exert the *greenhouse effect* which raises T to about 14°C. Again, anything that alters the greenhouse by changing the gaseous composition of the air may have a major effect on T. Today, the minor constituents of the atmosphere, carbon dioxide and water vapour, are chiefly responsible for the greenhouse effect; Sagan and Mullen (1972) have suggested that ammonia vapour acted in a similar fashion during the earlier history of the Earth, when the Sun emitted less energy than it does now.

Atmospheric particles, including dust from volcanoes and wind-blown soil, constitute a third influence in the interchange of energy between Sun, Earth and space. They may act exactly like dirt on greenhouse windows, and tend to lower T.

Ignoring possible continuing changes in the Sun, though, it is hard to see why T should change, as long as the albedo, the greenhouse and the particle-content are constant. Indeed, for 64 of the past 65 million years, the *mid-latitude* mean temperature (not T itself) remained close to 18°C. But during the past million years or so, the temperature has fluctuated widely, giving rise to repeated ice ages and warmer periods in between, like the one we live in. The most recent swings have been between the present mid-latitude value of 10°C and a low of −2°C, and their cause remains unknown. The asterisks in the diagram indicate entries which have served one expert or another as 'explanations' of the ice ages. That such diverse phenomena can be singled out as *the* principal cause confirms that we just do not understand the system.

The double-headed arrows indicate reciprocal influences and very strong coupling. Ocean currents, for instance, are wind-driven, but they transport the heat which drives the wind. Any closed loop of arrows beginning and ending on T has the potential of showing feedback. Some of this feedback has been investigated and found to be strongly positive, or reinforcing. For example, an increase of sea ice increases the albedo, reducing T and permitting more sea ice to form. Such positive feedback makes it seem difficult for the Earth to get out of an ice age once in; yet it has recovered several times in the past million years, with fluctuations so regular that they practically demand an underlying mechanism.

The circled question mark in the diagram does not stand for unknown parts of the system, but for the conceivable 'intransivity' of climate. Some experimental and theoretical approaches suggest that several stable values of T may be possible in any given set of environmental conditions, and that the choice between them is merely accidental. Thus it may be that climate is fundamentally indeterminate, with long-period changes in T arising quite at random.

My diagram is essentially a condensation of *Inadvertent Climate Modification* (SMIC 1971), to which the reader is referred for further discussion and such quantification as the data presently permit. The interim conclusion is that, although the diagram already excludes many interactions and minor effects, we are unable to simplify it further without a great deal of international and interdisciplinary effort in observation, experiment and theory.

In the midst of this uncertainty, there is reason to think that man's inputs – some of which are shown peripherally in the diagram – are no longer insignificant compared with the natural processes. New York City, for instance, excretes nearly seven times as much non-solar heat as it receives from the Sun, creating for itself an urban microclimate quite different from its surroundings, while smoke from automobiles and chimneys, and dust from eroded soil, affect the planetary abundance of atmospheric particles. The global temperature system is clearly one which is neither stable nor understood. It behoves man to be extraordinarily circumspect about tinkering with such a system, lest we find that we have perpetrated a truly global ecoblunder.

Complexity and man

In man's hunting phase, the human population was a regulated system, kept in steady-state by the same predator-prey-parasite interactions and environmental pressures as operate among other species. Man moved into 'almost-steady-state' after the invention of agriculture, with his population growing slowly and steadily. The Black Death of the fourteenth century was the last instance of negative feedback acting forcibly on the greater part of mankind, especially in regions of high population density. Since then the human population has been in the 'no-known-regulator' category. Many painful forms of negative feedback have been visualized for the near future. Yet, if man were to 'enjoy' the technological 'success' of complete environmental control, regulation would ultimately come only from the difficulties of radiating waste heat into space.

J. H. Fremlin (1964) sees this fundamental limit being reached with a human population of about 60,000 million million people, or 120 per square metre. By that time the roof that man will have to put around the whole planet will be at a temperature of the order of 1,000°C. At recent rates of population growth we shall reach that condition in less than 1,000 years.

There is no entry in our table of global regulators which a 'technological optimist' would place beyond man's capacity to control. But before we replace them with 'more efficient' systems designed by engineers, we should perhaps ask if they are significant to man in other than purely physical ways. A thought-provoking clue lies in the common features of the systems we have examined: an unsuspected labyrinth of causes behind an apparently straightforward consequence. This is quite unlike the situation in mechanics, whose simple linear behaviour has encouraged technologists to apply equally simple systems to the real world. This approach may be adequate for mere physical survival, but it is not at all clear that we can survive *as humans* in a world simplified by the engineer. What if a need for the complexity of nature is woven into our genes as an essential part of our humanity?

The intricacy of natural systems need not preclude man from working with them. The incredible tangle of civil law whereby man regulates human interactions shows that we are not totally inept at involved matters. A symphony is no less complexly structured than, say, upper-atmosphere chemistry. But in our approach to the natural environment we have so far been overly simplistic: written by engineers, conducted by economists, and performed by bureaucrats, our attempts to regulate nature have shown all the finesse of a tone-deaf child playing the Ninth Symphony on a kazoo.

References and further reading

J. H. Fremlin, 'How many people can the world support?', *New Scientist*, 24 (1964), pp. 285-7.

H. Johnston, 'Reduction of stratospheric ozone by nitrogen oxide catalysts from supersonic transport exhaust', *Science*, 173 (1971), pp. 517-22.

F. MacIntyre, 'Why the sea is salt', *Scientific American*, 223 (1970), No. 5, pp. 104-15.

C. Sagan and G. Mullen, 'Earth and Mars: evolution of atmosphere and surface temperatures', *Science*, 177 (1972), pp. 52-6.

SCEP (Study of Critical Environmental Problems), *Man's Impact on the Global Environment* (Cambridge, Mass. 1970).

G. G. Simpson, *Tempo and Mode in Evolution* (New York 1944).

SMIC (Study of Man's Impact on Climate), *Inadvertent Climate Modification* (Cambridge, Mass. 1971).

W. L. Thomas, Jr (ed.), *Man's Role in Changing the Face of the Earth* (Chicago 1956).

THE CURRENCY
OF LIFE
Park S. Nobel

Plants and animals require a continual input of energy. If we were to remove the source of energy, organisms would drift toward equilibrium and consequent cessation of life. This chapter is concerned with energy which is available for performing biological work, and which thermodynamicists have called the *free energy*. Photosynthesis converts the plentiful radiant energy of the Sun into free energy stored first in intermediate energy 'currencies' like ATP (adenosine triphosphate) and then in the altered chemical bonds which result when carbon dioxide and water react to form carbohydrate and oxygen. Thereafter, as plants grow and animals feed on them, the free energy represented by carbohydrate is reconverted by respiration into other energy currencies which can do the necessary work – whether it be to amass salts and amino acids in a cell, to pump blood, to lift weights, or to power the electrical machinery of the human brain.

These processes, occurring on both a molecular and a global scale, are the very essence of life on Earth. At every step, whether it be a chemical reaction or one organism eating another, the amount of free energy decreases. The biological environment is therefore, to the thermodynamicist, an arena in which free energy gradually dissipates itself in a long chain of events. The structure of the

Park S. Nobel is associate professor of molecular biology and a member of the Institute of Evolutionary and Environmental Biology at the University of California, Los Angeles. His research, writings, and teaching are primarily concerned with the physiology and energetics of plants.

ecosystem and the quantity and quality of life it can support are controlled by the initial supply of free energy and by the inexorable laws of thermodynamics that govern the 'expenditure' of that free energy. So fundamental and so widely applicable is the idea of free-energy conversion that we will next examine it in some detail.

Concept of free energy

Suppose that a living system (a cell, a tissue, or entire plant or animal) changes its state in some way. Something happens to it, a plant cell makes some protein, an animal contracts a muscle, or any one of many other agents or mechanisms operates. If work has to be done to go from state A to state B, the change in free energy is equal to the *minimum* amount of work needed. If, on the other hand, work can be obtained from the transition from A to B, the free-energy change is the *maximum* amount of work which can be derived. Thus, the concept of free energy is concerned with limits to the work done on or by a system when it undergoes a change.

Most biological systems are subject to a constant (atmospheric) pressure and remain at constant temperature, at least for short periods. This is quite different from the situation in, say, a car engine. For discussing the energetics of processes at constant temperature and pressure, the appropriate quantity is known as the *Gibbs free energy*, and throughout the rest of this chapter 'free energy' refers to this form. To every chemical compound in a given state a measure of its free energy can be assigned and this quantity is called its *chemical potential*. Chemical potentials are often expressed in kilocalories (or kcal) per mole. A mole constitutes a particular number of molecules of the substance in question – Avogadro's number, the number of molecules in 32 grams of gaseous oxygen, or roughly 6×10^{23} (6 followed by 23 zeros).

Any change in the system which reduces the chemical potential of some chemical compound can occur spontaneously. No outside source of free energy is needed; on the contrary, work can be done by the system. An example is water flowing downward. A change which increases the chemical potential of a substance can occur only if some other change occurs in the system, this other change supplying the free energy required. The pumping of blood along arteries is a case in point. Finally, if no change in chemical potential of some chemical compound occurs in the change from A to B, the material in question is at equilibrium. Living systems as a whole are usually far removed from the equilibrium condition.

We are often concerned with the free-energy change for a mixture of substances engaged in a chemical reaction. The chemical potential of each reactant and each product enters into the calculation; each chemical potential is simply multiplied by the quantity of the substance in question (reckoned in moles) to give its contribution to the energetics of the system. The change in free energy for a chemical reaction is then the total free energy for the products *minus* the total for the reactants. If this is a negative quantity, meaning that the free energy decreases, the reaction can take place spontaneously. This is the case for the reaction of carbohydrate and oxygen to form carbon dioxide and water. When the calculation indicates that the free energy must increase during the reaction, then a free-energy input at least equal to that increase is needed in order to convert the reactants into products. When a plant is forming carbohydrate and oxygen from carbon dioxide and water, sunlight provides the energy input.

The free energy of a substance plainly depends on its chemical composition, but it is also influenced by other factors. For example, water in the leaves of a tree, by virtue of its height above the ground, has additional potential energy compared with water in the roots. Materials can also possess electrical potential energy if they carry an electric charge. In intuitively less obvious ways, the free energy of a material also depends on its concentration and on the prevailing pressure. All of these factors have to be taken into account, in a quantitative fashion, in assessing the chemical potential of a particular component, as is summarized in the table overleaf.

Some of the types of biological work that can be accomplished by 'spending' free energy help to illustrate the various terms making up the chemical potential. At the same time, we can note the versatility of one of the energy-rich materials used as currency – ATP. For the operation of the machinery of a living cell, the ATP molecule is an exceptionally convenient form for short-term storage and distribution of free energy.

One of the characteristics of living cells is the ability to maintain the internal concentrations of certain dissolved neutral materials, like glucose or the amino acid glycine, far above their concentrations in the surrounding water. As we have indicated, the chemical potential of such a substance depends on its concentration, which is

---

Contributors to the chemical potential of a substance 'j'
(For the derivation, see Nobel 1974)

---

chemical	=	constant	+	concentration term	+	pressure term	+	electrical term	+	gravitational term

$$\mu_j \;=\; \mu_j{}^* \;+\; RT \ln a_j \;+\; V_j P \;+\; z_j FE \;+\; m_j gh$$

Key:

$\mu_j{}^*$	chemical potential of j in an arbitrary standard state
R	universal gas constant
T	prevailing absolute temperature
$\ln a_j$	logarithm of the effective concentration of j
V_j	volume per mole of j
P	prevailing pressure
$z_j F$	charge per mole of j
E	prevailing electrical potential
m_j	mass per mole of j
g	acceleration due to gravity
h	prevailing height above an arbitrary level

---

often higher inside the cell than outside. To transfer a mole of a neutral compound in the direction of a tenfold increase in concentration requires 1.4 kcal of free energy (at 30°C), while it takes 2.8 kcal for a 100-fold and 4.2 kcal for a 1,000-fold increase in concentration. ATP can supply about 12 kcal per mole and it often serves as the free-energy source for what is called 'active transport' of materials across membranes toward higher chemical potentials.

Specialized cells, of which animal nerves are the most familiar, transmit electrical impulses along their length for purposes of communication. The impulses involve the movement of sodium and potassium ions through the cell membrane, which often changes the electrical potential difference across the membrane by about a tenth of a volt. To return the cell to its original condition, in preparation for the next impulse, can then require the active transport of ions. If the concentrations of each ion are the same on the two sides of the membrane, a free energy input of 2.3 kcal is required to transport a mole of potassium across an increase in electrical potential of a tenth of a volt. Again, the currency supplying the free energy can be ATP.

ATP is also the free-energy source for the contraction of muscles. The mechanical energy of contracting muscles can be used to propel fluids along tubes or to lift weights. For instance, the ATP-driven contraction of the muscles surrounding the left ventricle of the human heart can increase the blood pressure within it by 0.2 atmosphere, before the blood is released through the aorta. The circulatory system thus uses the free energy of ATP to cause muscle contraction which in turn increases the local pressure and hence the chemical potential of blood. Pressure-driven flow is actually quite an efficient way to move fluids, since it takes only 0.005 kcal of free energy to increase the pressure of one litre of blood by 0.2 atmosphere. As an example of gravitational work resulting from ATP as an energy currency, the contractions of the muscles in our legs allow us to climb hills. The increase in free energy as a 50-kilogram person ascends 100 metres is 12 kcal.

Once the concept of free energy has been grasped, it becomes a powerful aid to understanding processes in the biosphere. Here we will begin to use it more systematically, first in considering the initial supply of free energy.

Sunlight and photosynthesis

Through a series of nuclear reactions taking place within the Sun, nuclear mass is converted into energy in accordance with Einstein's famous relation, $E = mc^2$. Of the large amount of energy that the Sun radiates into space, about 1.3×10^{21} kcal annually falls on the Earth's surface. Approximately 5 per cent of this radiant energy, or 6×10^{19} kcal, is absorbed by chlorophyll or other photosynthetic pigments in plants. This absorption initiates the conversion of solar energy into various forms of chemical energy by photosynthesis. Such chemical energy may then be stored in the plants, mainly in the chemical-bond changes involved in the synthesis of carbohydrates. Plants act as the direct source of free energy for herbivorous animals and indirectly serve as the source for carnivores which eat the herbivores.

About half of the solar radiation reaching the Earth's surface occurs in the visible region of the spectrum, where it can be absorbed by the three types of photosynthetic pigments (see the diagram overleaf). Of these pigments, chlorophyll is essential for photosynthesis in all plants and it absorbs blue and red light – reflecting the green, which gives the rural landscape its characteristic colour. Carotenoids absorb blue and some green light and pass their

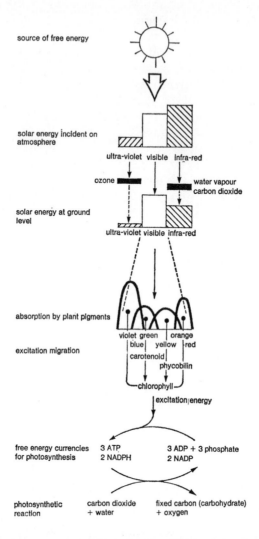

The source of free energy for life is the Sun. Much of the ultra-violet and infra-red radiation is absorbed in the atmosphere and so most of solar energy reaching the ground is in the visible region of the spectrum. Pigments in plants absorb the light energy, which is then converted into chemical potential energy in the form of ATP and NADPH. These two free energy currencies drive the process of photosynthesis energetically 'uphill'

excitations on to chlorophyll; phycobilins, absorbing primarily in the yellow and the orange regions, act similarly in certain algae. This absorption of sunlight by the photosynthetic apparatus of plants and some bacteria sets the stage for the conversion of the radiant energy into chemical forms that are needed for life.

Soon after the trapping of radiant energy by photosynthetic pigments, free energy is stored in ATP and in NADPH. As mentioned earlier, ATP is adenosine triphosphate, while NADPH is the reduced form of nicotinamide adenine dinucleotide phosphate, NADP. Both of these compounds act as energy currencies for the chemical reactions of photosynthesis, supplying the energy needed to convert carbon dioxide and water into carbohydrate and oxygen. More generally, they represent the two main classes of short-term energy-storage compounds used by all plants and animals.

ATP is made from ADP (adenosine diphosphate) in a crucially important reaction:

$$ADP + \text{phosphate} \rightleftharpoons ATP + \text{water}$$

The double arrow indicates that the reaction is reversible. In the conditions that generally occur in a cell, a free-energy *input* of about 12 kcal per mole is necessary in order to form ATP. Thus, the reaction forming ATP does not proceed spontaneously but must be coupled to an energy source – in this case, the light absorbed by photosynthetic pigments. In the reverse reaction ATP is hydrolyzed back to ADP and phosphate, which releases about 12 kcal of free energy per mole of ATP.

In photosynthesis three moles of ATP are generally involved in the conversion of one mole of carbon dioxide into carbohydrate; two moles of NADPH are also needed. The electrons that reduce NADP to NADPH come from water with the concomitant release of oxygen, a part of photosynthesis that is rather poorly understood. However, we do know how much free energy is involved: an input of approximately 53 kcal per mole is necessary to reduce NADP to NADPH with the evolution of a half mole of oxygen from water.

We now return to a consideration of sunlight and of how efficiently the photons of which it consists are used in forming ATP and NADPH. A convenient unit for light energy is that of a mole (Avogadro's number) of photons. A mole of blue photons has about 60 kcal of energy while a mole of red photons corresponds to approximately 40 kcal (the exact amounts depend on the wavelength), so that photosynthetically active visible light has on the average about 50 kcal per mole.

Approximately eight moles of such photons, corresponding to about 400 kcal of energy, are involved in producing the three moles of ATP and two moles of NADPH which in turn are needed for fixing one mole of carbon dioxide. Thus the light-energy input of 400 kcal leads to a free-energy storage of $(3 \times 12) + (2 \times 53)$ or 142 kcal in the two energy currencies. By the first law of thermodynamics (the conservation of energy principle) energy can neither be created nor destroyed but only converted into other forms. In the present case the 'missing' 258 kcal are dissipated as heat which, at constant environmental temperature and pressure, cannot do biologically useful work. The efficiency of the initial conversion of light energy into chemical energy is therefore 36 per cent. This figure relates to light energy actually absorbed by chlorophyll and other photosynthetic pigments under optimal conditions, which is only a part of the sunlight falling on a plant.

Let us follow the process to the next stage, the formation of carbohydrate, typically glucose, from carbon dioxide. The free-energy change in fixing six moles of carbon dioxide into one mole of glucose is 686 kcal; in other words, 114 kcal are stored per mole of carbon dioxide fixed. The three moles of ATP and two of NADPH which achieve this feat bring with them 142 kcal, as we indicated above. Thus 80 per cent of the free energy of ATP and NADPH is retained in making the carbohydrate. Evolution has indeed led to an extremely efficient set of chemical reactions for photosynthesis. It is impossible to attain 100 per cent efficiency for free-energy retention, in consequence of the second law of thermodynamics, which can in fact be paraphrased as follows: 'The free energy decreases for each set of chemical reactions that occurs'.

We can now consider the fate of the free energy stored by photosynthesis in the carbohydrate glucose. When used as a fuel, glucose in principle simply reacts with oxygen to make carbon dioxide and water, reversing the process of its formation and releasing energy. In practice, the course of events is subtler. It includes, for example, the reduction of NAD (nicotinamide adenine dinucleotide) to NADH; this is yet another energy currency, similar to NADPH. In turn, the NADH is reconverted in a process that produces ATP from ADP and phosphate. The controlled combustion of one mole of glucose leads to the net formation of 38 moles of ATP; the overall efficiency of this many-faceted process is 66 per cent. Thus the free energy in glucose can be efficiently mobilized to produce ATP, a short-term energy currency. For long-term storage of free energy, glucose is generally incorporated into polymers like starch and glycogen or else metabolically used to form fat.

Energy flow through the biosphere

We have followed the flow of energy from the Sun to intermediates like ATP and NADPH to carbohydrates and then back to ATP for use by cells. Having established the concept of free energy at a molecular level, the thermodynamicist can apply the same principles to environmental systems. As the free energy flows between plants, micro-organisms and animals, the energetic considerations can become very complex. The dominant rule – the second law of thermodynamics – still operates and we lose free energy at each step along the way.

The harnessing of solar radiation by photosynthesis in plants starts the flow of free energy through the biosphere. In addition to maintaining individual chemical reactions as well as entire plants and animals in a state far from equilibrium, the annual degradation of chemical potential energy to heat sustains the great chemical cycles of the biosphere: of oxygen, released in photosynthesis and then consumed by respiration; of carbon, cycling in the reverse direction between these two processes; of nitrogen, taken up during growth and released in decomposition; and so on. The repeated formation and break-up of energy-rich ATP is a molecular cycle, mimicking the cycle of life and death of organisms. All of these material cycles are a consequence of the unidirectional flow of ever-diminishing free energy.

The energy storage by photosynthesis can be evaluated on a global scale. We have already noted that chlorophyll and the other pigments involved in photosynthesis absorb about 6×10^{19} kcal per year of solar energy, and that 400 kcal of radiant energy are convertible to 114 kcal of free energy in carbohydrate and oxygen formed from carbon dioxide and water. Under ideal conditions of temperature, illumination and physiological status of the plants, the absorbed sunlight could make carbohydrates representing about 2×10^{19} kcal of free energy per year. Under natural conditions averaged over a year, however, the conversion of radiant energy to free energy is much less – about 6×10^{17} kcal or about one per cent of the absorbed energy. This still represents an enormous amount of carbohydrate – about 150,000 million tons per year – much of it made by plankton living near the surface of the oceans. Such a large annual production is difficult to visualize, but it would, for example, correspond to a giant tree of solid carbohydrate a kilometre in diameter and about two hundred kilometres high. Of course, this is not the actual growth of plants, because much of the carbohydrate is consumed as fuel almost as soon as it is made. Even so, the 6×10^{17} kcal of free energy that it represents is the initial supply of the free energy used annually by all plants and animals.

About 60 per cent of this free energy is dissipated by plants through their own respiration. The herbivores that graze on the plants generally retain only 10 to 20 per cent of the free energy of the ingested material, in the case of growing animals; a mature animal uses essentially all of its free-energy consumption just to remain as it is, in a state far from equilibrium. Growing carnivores feeding on herbivores can store about 10 to 20 per cent of the free-energy content of the meat. Similarly 'top carnivores' that feed on carnivores can, at best, retain up to 10 to 20 per cent of their free energy. Thus, there is a sizable loss in free energy for each link in a food chain, which as a consequence seldom has more than four steps: plants, herbivores, carnivores, top carnivores.

Although man can shorten the length of some of his food chains and thereby minimize the loss of free energy, he nevertheless makes a rather large demand on the free energy available in the biosphere. The global average intake of free energy in food is 2,400 kcal per person per day; multiplied by the world population in 1973, this corresponds to 4×10^{15} kcal per year. Man also consumes plants and animals for clothing, shelter, firewood, papermaking and many other purposes. Leaving these demands aside, man's food consumption alone amounts to almost one per cent of the free energy stored by all of the plants. It is instructive to reflect on what would happen if man lived as a top carnivore, fancying the meat of cats and sharks. With a 10 per cent retention of free energy at each link in the food chain, he would then be indirectly responsible for the consumption of the *entire* storage of free energy by present-day photosynthesis.

Fortunately, man is omnivorous and he obtains most of his free energy from plants. For instance, the average daily consumption in the United States is 3,100 kcal per person, of which 2,200 kcal comes from plants and 900 kcal from animals. In turn, the animals consumed require an average free energy input of 6,300 kcal of plant material in order to produce the 900 kcal, which represents a 14 per cent efficiency for free-energy retention at this link in the food chain. If we imagine a world population twice as large as now, fed to present-day United States standards, human food consumption would represent 5 per cent of the total free-energy supply for all life on Earth. Were it not for advances in agriculture which help both to increase the total supply of free energy and to sequester it for human use, our species would already be losing its free-energy credibility.

References and further reading

A. L. Lehninger, *Biochemistry* (New York 1970).
H. J. Morowitz, *Energy Flow in Biology* (New York 1968).
P. S. Nobel, *Plant Cell Physiology: A Physiochemical Approach* (San Francisco 1970).
P. S. Nobel, *Biophysical Plant Physiology* (San Francisco 1974).
Scientific American (an issue devoted to energy), 224 (September 1971), 3.

THE THIRSTY SOIL
J. S. G. McCulloch

Most environmental systems are complex, comprising as they do a whole series of processes which interact in diverse ways. Sometimes this interaction is additive: air temperature at a given place and time may be considered to be the *sum* of the many contributing processes such as radiation, wind on scales varying from the very small to the hemispherical, altitude, vegetation and the nature and wetness of the underlying ground. Some of these tend to increase and others to decrease the temperature but none is overriding; even an eclipse of the Sun does not send the temperature plummeting to absolute zero. The growth of a plant is also subject to many influences but in a different way. Unlike the temperature, the rate of growth can easily fall to zero when supplies of light, water or any of the various major and minor nutrients are lacking. These factors are in effect *multiplied* together and setting any one of them at zero may reduce the rate of growth to zero. In such circumstances it becomes possible to speak of limiting factors.

In environmental science the concept of limiting factors helps in simplifying the analysis of a complex system and in evolving policies for improving the health of the system. In extreme environments the limiting factors are conspicuous: no amount of fertilizer would assist in farming the deep Sahara without water; nor would warmth and water at the South Pole in winter induce plants to grow without light.

J. S. G. McCulloch, director of the Institute of Hydrology of the (UK) Natural Environment Research Council, was formerly head of the Physics Division of the East African Agriculture and Forestry Research Organization.

In ordinary conditions the concept of limiting factors is by no means so straightforward. Components of the system can be limiting by excess as well as by deficiency – there can be too much of a good thing, whether it be scorching sunlight, water in flood or an essential nutrient such as copper which becomes a poison at a certain concentration. Organisms have a range of tolerance for each factor and a species may be more sensitive to one factor than to another. As conditions change, first one factor and then another may be limiting. If one limit is removed, other limits will come into play; for example the benefit of a shower of rain or of irrigation by man may be nullified by a lack of nitrogen in the soil. Add to these considerations the fact that two or more factors are often coupled physiologically and that organisms respond to adverse conditions in ingenious ways, and it becomes apparent that the concept of limiting factors is not a sword that will infallibly cut the Gordian knot of environmental complexity.

Nevertheless it remains, in many cases, the best hope for evolving practical policies for improving the wellbeing of an environmental system including man. Whether or not we are always successful in identifying them, limiting factors operate in all living systems. The concept is also helpful in considering non-living systems including technological processes and also the human economy at large which may be subject to limits set by shortages of strategic minerals and the like. Although it may be seldom obvious which of many variables is the limiting factor of primary importance, in any one environment human experience – the outcome of many years of trial and error – gives guidelines for sound management. Such empirical knowledge is not transferable to a different environment without pilot studies to test its relevance.

The extension of the Indian tea-shade tradition to Kenya is a pertinent example of unwarranted extrapolation from one environment to another. The long-established practice in Assam of planting shade trees throughout tea gardens was the subject of scientific investigation by the Tochlai Experimental Station of the Indian Tea Association. Tea yields under shade were found to be greater than those obtained without shade (Wight 1958) – a result in accord with growers' experience of tea and of other tropical crops such as cocoa. The shade tree remained rooted in the accepted management of tea gardens. With development of tea estates in Africa, and particularly at high altitude in Kenya, shade trees were established throughout the gardens. However, investigation in Kenya by the East African Agriculture and Forestry Research Organization produced the surprising result that tea yield was *inversely* proportional to shade amount (McCulloch 1965).

This apparent contradiction in the results of well conducted but *ad hoc* experimentation shows the need for a complete understanding of the process and of local conditions to explain the empirical results. Further work at Tochlai finally provided the explanation: in Assam intense sunlight was associated with high air temperatures and hence, for the tea plant, excessively high leaf temperatures; in Kenya equivalent or even higher levels of isolation resulted, at the high altitude of the tea gardens, in moderate air temperatures. In Assam high leaf temperatures were the limiting factor of primary importance, and the shade trees reduced them. In Kenya the shade trees simply competed with the tea for soil moisture, there the dominant limiting factor, although they sheltered parts of the tea gardens from the intense hail storms which occasionally occur in the tea-growing region of Kenya.

This case illustrates the interactions between various factors. Careful research in realistic conditions is needed for disentangling them and the time taken for plants to grow and the variations in weather from year to year mean that the research cannot be hurried. Moreover it must be open-minded, not only about conventional land-management practices but also about the possible kinds of limiting factors; nature pays scant regard to the conveniences of scientific specialization. As mankind has known since his origins, water availability is a major limiting factor in many parts of the world. Yet the hydrologist, whose concern is with the science of water, must be aware of the many other types of limiting factors which can influence the hydrological regime in any area under study. My second example, also from Kenya, shows how a non-hydrological factor turned out to have important consequences for the management of water.

In East Africa annual rainfall exceeds the annual potential evaporation only on 4 per cent of the land surface and the harvest of water from the relatively small high-altitude high-rainfall catchment areas is of vital importance to a population growing both in numbers and in sophistication. There are pressing local requirements for information on the hydrological consequences of man's interference with his environment by changing the land use. In the

Aberdares above Nairobi, for example, can retention of unprofitable indigenous bamboo be justified in order to safeguard water supplies, when crops of vegetables or coniferous softwoods are more economic? At Kericho in the South West Mau Forest Reserve, can tea gardens replace uneconomic montane rain forest without increasing losses by evaporation or causing adverse effects on the quantity, quality and distribution of river flow from the area?

Where rainfall is frequent and plentiful, no great storage of moisture in the soil is necessary and the root development of plants tends to be restricted to a shallow surface layer of not more than 1 metre depth. In the balance between rainfall, evaporation and stream runoff, changes in storage and percolation to ground water are generally minor terms and hydrological studies in wet temperate regions reduce almost to statistical sampling of rainfall input followed by hydraulic studies of flow output. Where rainfall is strongly seasonal with long dry periods, as in parts of East Africa, plants exploit deep, well-drained soils to 3 metres and, in the case of perennial grassland or forest, even to 10 to 15 metres. Under the latter conditions, which prevail in the catchment areas, the storage of moisture in the soil becomes a major component of the hydrological cycle. The process of exchange between the components of the hydrological cycle can be investigated more precisely in areas where each of the components is of a similar order of magnitude.

In 1957 two of the most valuable hydrological investigations of recent years began in Kenya as part of a complex of four large-scale experiments in catchment areas, in Kenya, Uganda and Tanzania. The initial results (Pereira 1962) have subsequently been augmented (Dagg and Blackie 1965; Blackie 1970). The techniques of experimentation were straightforward but more thorough than most previous investigations. Besides the usual record-keeping of rainfall, streamflow and the meteorological factors which control evaporation, measurements were also made of soil moisture storage. By successive approximations the water balance equation was solved for the soil moisture storage and the results were compared with the direct measurements. These confirmed the scientific basis of the approach and highlighted the need for more detailed studies of the processes in the hydrological cycle so that the immediate results of these and similar investigations might be extrapolated to different environments.

What then was the bearing of these experiments on land use?

Previous work (Penman 1948) had shown that evaporation from short crops was determined by the climate rather than by the character of the crop. The East African studies suggested that this result was also generally applicable to taller vegetation, irrespective of whether it was 12-metre-high bamboo, 6-metre-high pines or 1-metre-high tea gardens interplanted with shade. However the upheaval caused during the change in land use did result in a marked although temporary deterioration in the ability of the catchments to retain moisture. During the transition period the river regime became flashier, in that flood peaks were higher and of shorter duration. Finally the quality of streamflow under flood conditions deteriorated sharply, particularly in catchments where the new land use was labour-intensive, except where extensive, and expensive, soil conservation measures were taken. Otherwise the literal pressure of man and animals in a small area resulted in compaction of the soil surface, accelerated erosion and flood runoff orders of magnitude greater than from the original forest cover.

Thus a limiting factor emerged: the population density of men and animals. The conclusion as far as management was concerned was clear. Catchment areas of importance for water supplies are best preserved undisturbed, in this region of Africa at least. If some more economic form of land use is essential then it should be chosen to minimize the amount of labour necessary unless substantial investment in soil conservation works and in impounding structures for water can be justified. These results and recommendations have been applied to several extensive areas in the Aberdares. Unfortunately, in one particular area, socio-economic pressure was so great that the indigenous forest was replaced by agricultural smallholdings. Although the hydrological limits for a balanced land use for the area in the long term were known and understood, short-term political pressure resulted in these limits being exceeded, knowingly.

There are, however, occasions when similar decisions are taken without knowledge of the possible consequences. The Mbarali irrigation project in Tanzania, funded by international capital, provides a salutary warning of the consequences of blind application of technology to an environment which has not been examined in sufficient detail. Hydrologists, particularly in developing countries, consider themselves fortunate if they have a moderately long series of years of daily observations of 'stage', the river level at a suitably stable section of river. The relationship between stage and volume of flow, empirically derived, enables the appropriate hydrological statistics to be extracted. The most important of these are predictions of the frequency of occurrence of floods of given magnitudes, on which hydraulic designs of works are based. Before the establishment of the Mbarali irrigation scheme, river records for a tributary of the Great Ruaha, the proposed source of irrigation water, were examined on the assumption that the physical condition of the contributing catchment area had not altered over the period of the records. In fact the relevant catchment area, in the Kipengere Range, had suffered massive deterioration through years of overgrazing and uncontrolled peasant cultivation in inappropriate areas. In consequence the soil moisture storage of the catchment was reduced substantially and its hydrological behaviour was much more erratic than the statistics suggested. At the first major flood the water intake works for the irrigation scheme were washed away.

Hence in man's attempt to adapt his environment to his own convenience he is everywhere faced by factors which may well prove limiting. If the hydrology is well understood, the value judgements correct and the technology adequate, then designs of works calculated to alter the hydrological cycle will be conservative and will take account of the limiting factors. Each and every phase of the cycle may in a particular environment prove limiting. Over much of Africa rainfall limits vegetation. In fertile areas man may overcome this limitation by irrigation, either by run-of-the-river schemes dependent on conservation of the headwaters catchment areas, or by massive dams or like structures to impound the necessary waters. The amount of water used by evaporation from different crops in a single environment is surprisingly similar, except of course at seed times and at harvest. Management practices can help to adapt a crop to the available moisture, but the limiting factor, the potential evaporation, is determined mainly by the climate. Man's impact on soil moisture storage tends to be negative rather than positive; changes in land use all tend to reduce rather than to increase the available storage.

The flow of water in rivers to the sea is what hydrology leads to and what hydraulics is all about. The safety of life and property in the flood plain of a river can be predicted in a statistical sense and hydraulic works may at great expense alleviate the danger in one area, perhaps to the detriment of another area. With increasing development and increasing population pressure, the trend of river regimes is towards more immediate response to rainfall and hence towards more violent floods. So the urban and industrial planner cannot afford to ignore hydrological changes, any more than the farmer can, while the hydrologist in turn must try to understand the non-hydrological factors that bear upon the systems he investigates.

Therefore let the role of the environmental scientist, whatever his specialist training and preoccupations, be one of drawing together the numerous factors which may be limiting in any particular situation. Environmental science should not be left as a set of over-simplified, discrete sciences among which the vital limiting factor may be overlooked. Each science must contribute its expertise in unravelling the individual processes but all must in the end be synthesized into the science of the environment.

References and further reading

J. R. Blackie, 'Hydrological effects of a change in land use from rain forest to tea plantation in Kenya', presented at symposium on the results of research on representative and experimental basins, International Association of Scientific Hydrology (Wellington, New Zealand 1970).

M. Dagg and J. R. Blackie, 'Studies of the effects of changes in land use on the hydrological cycle in East Africa by means of experimental catchments', *Bulletin of the International Association of Scientific Hydrology*, 10 (1965), 4, pp. 63–75.

J. S. G. McCulloch, H. C. Pereira, O. Kerfoot and N. A. Goodchild, 'Effect of shade trees on tea yields', *Agricultural Meteorology*, 2 (1965), pp. 385–99.

H. L. Penman, 'Natural evaporation from open water, bare soil and grass', *Proceedings of the Royal Society* (London), Series A (1948), 193, pp. 120–45.

H. C. Pereira et al., 'Hydrological effects of changes in land use in some East African catchment areas', *East African Agricultural and Forestry Journal* 27 (1962). Special Issue, pp. 1–131.

W. Wight, 'The shade tree tradition in tea gardens of northern India, I, The value of shade', Indian Tea Association, Science Department, Tochlai Experimental Station Annual Report (1958), pp. 75–97.

Experimental environments

NUTRIENTS
IN A LAKE
D. W. Schindler

Ecologists have by tradition gathered their information by careful observation of natural ecosystems. They wait passively for the systems to change, in response to the weather or other conditions; then, by correlating the natural variations in the ecosystems, they attempt to discover the causal relationships that govern the populations of component species and the overall state of the ecosystems. Great difficulty usually surrounds such studies, because of the complexity of natural ecosystems and the interaction of their many components. It has been nearly impossible, in many cases, to distinguish causal relationships from spurious correlations, since the usual scientific method of setting up hypotheses and testing them by vigorous experiment is seldom open to the field observer. He is unable to manipulate the ecosystem to find out how it works.

For this reason, many ecologists have turned to small-scale, controlled experiments in the laboratory, where manipulation becomes easy. But here the great drawback is lack of realism. Only highly simplified, artificial ecosystems may be dealt with, owing to the difficulty of working with large, active or delicate members of natural food webs. The outcome of applying the results of laboratory experiments to complex natural systems is therefore extremely

D. W. Schindler is in charge of the Experimental Lakes Area of the Fisheries Research Board of Canada. He is based at the Freshwater Institute, Winnipeg.

doubtful and any far-reaching predictions about natural ecosystems must certainly be made with extreme caution.

With recent public demands for proof of environmental damage and the causes of it, and for the planning of environmentally-sound industries, farms and human communities, a more powerful and speedier approach to environmental problems is required. We have to do experiments with real ecosystems, in which we can manipulate a single chosen element of a system, in the absence of any confusing sources of variation, and observe its effects on all other components of the system. This is the role of experimental environments, tracts of land or water selected as open-air laboratories.

The concept of the experimental environment is not a new one. Agriculture has employed such methods for decades, in the form of experimental plots and farms, where animal and plant communities of interest to the farmer are treated in controlled ways to discover factors influencing their development – for example, the effect of different fertilizer treatments on the yield of a cereal crop. Agricultural ecosystems, however, tend to be highly simplified, with their species composition dictated by man. This comparative simplicity is characteristic of most man-dominated environments, and interpretation of results is usually rather straightforward. In natural environments, where species are numerous, their relationships to each other obscure, and a single, clearcut management objective does not exist, devising experiments and interpreting their results are far more demanding.

Nevertheless, experimental approaches to natural systems can provide useful and unique insight into the functioning of an ecosystem. As an example, I shall discuss studies in the Experimental Lakes Area (ELA) of the Fisheries Research Board of Canada, a cluster of small, natural lakes in north-western Ontario taken over for research. These studies have played a useful part in deciphering causes of eutrophication, the excessive growth of aquatic plants – the phytoplankton – which is injuring many lakes by exhausting their oxygen.

Eutrophication in Lake 227

The background to this research is the recent controversy over whether phosphorus, carbon, or both are responsible for the eutrophication of North American freshwater lakes, including the

Great Lakes that divide Canada from the United States (Likens 1972). For years, freshwater biologists have believed that phosphorus supplied by sewage, detergents and agricultural fertilizers is the cause of cultural eutrophication. Phosphorus was identified as the key element controlling phytoplankton growth in the majority of observational field studies (for example, see Hutchinson 1957) and also in experimental laboratory work with algal cultures (for example, see Chu 1943).

In spite of several thousand relevant papers in the scientific literature, the evidence was not sufficient to convince North American industrialists and politicians that they should spend millions of dollars for the control of phosphorus. A small but vociferous group of investigators claimed that carbon, not phosphorus, was responsible for eutrophication. This contention, based on a small number of laboratory studies, was able to cloud the issue sufficiently to delay phosphorus-control legislation for several years. This delay has undoubtedly cost the taxpayers of North America tens of millions of dollars a year in lost resources. Volumes of careful observational work failed to give a definitive picture of the role of phosphorus in eutrophication. There always seemed to be a 'loophole', either in gaps in the prodigious amount of data which had to be collected to examine one topic, or else in the analysis of all the influences and modifications of various sorts that affect a lake simultaneously. On the other hand, no industrialist appeared willing to risk his corporate profits, nor any politician his office, on evidence extrapolated from small laboratory experiments done in flasks or test tubes. The problem of supplying facts to resolve this controversy appeared to be ready-made for an experimental environment.

We set up an experiment in Lake 227 as an extreme test of the carbon hypothesis. Preliminary studies had revealed that this lake was lower in dissolved inorganic carbon (carbon dioxide + bicarbonate + carbonate) than any previously studied natural lake; indeed, special gas-chromatographic techniques were needed to detect any dissolved inorganic carbon at all. Beginning in 1969, we added phosphorus (as phosphate) and nitrogen (as nitrate) to the lake in quantities similar to those reaching Lakes Ontario and Erie. If Lake 227 eutrophied under such circumstances, the carbon-limitation theory could hardly be of any practical significance, since all lakes of economic importance contain and receive far more carbon than Lake 227 does. Except for this peculiar lack of carbon, the waters of our study lake were very similar to those from other areas of the Canadian Shield. Since the Shield supplies more than half the water to the St Lawrence Great Lakes, and contains at least 80 per cent of all the world's lakes, any inferences we could draw would be correspondingly relevant to huge areas of fresh water.

Lake 227 began to show symptoms of eutrophication within a few weeks after phosphorus and nitrogen additions began, in spite of the apparent shortage of carbon. By the end of summer, in 1969, phytoplankton had increased at least twenty-fold and, by August 1970, we measured blooms of nearly one hundred times the prefertilization concentrations. The small yellow-brown algae (*Chrysophyceae* and *Crpytophyceae*) which characterize small Shield lakes had given way to blue-green algae of the genera *Oscillatoria, Pseudoanabaena,* and *Lyngbya* (see Schindler *et al.* 1971).

A number of important observations were made during the first three years of fertilization. First, there was no increase in dissolved phosphate in the lake, in spite of the fact that we were adding phosphorus at a rate of ten micrograms per litre per week. Using traces of radioactive phosphate we were able to show that the plankton took up virtually all of the added element within minutes after fertilization. Nitrogen, for which planktonic demand was lower, was not depleted for three or four days after fertilization. The concentration of phosphate-phosphorus left in solution was nearly always less than one microgram per litre, far short of the ten micrograms per litre previously thought to be the critical concentration needed for provoking algal blooms. Thus our results contradicted the commonly believed 'Sawyer's Limits', upon which many lake-management schemes have been based.

Numerous measurements on components of the carbon cycle revealed that no carbon shortage actually occurred until the lake was highly eutrophic. The percentage of carbon in plankton of the lake remained identical to prefertilization values, and to values for unfertilized lakes in the area. Given the observed hundred-fold increase in the amount of plankton, this meant that an enormous amount of carbon was being supplied from somewhere to maintain the percentage in plankton. Since no carbon was added as fertilizer and very little entered either with inflows or from sediments in the lake, the atmosphere appeared to be the only possible source. This was confirmed by measurements of gas exchange between water and atmosphere made in 1970 and 1971, using radioactive carbon and a chemically inert radioactive gas, radon, and by total carbon budgets for the lake. There appears to be little if any possibility of a carbon shortage occurring in freshwater lakes, since the atmosphere represents an enormous reservoir able to support the largest phytoplankton bloom. The phantom of carbon-limitation can now be regarded as dead.

Other experiments

Lake 227 has provided a useful place for other critical ecological experiments, since its nutrient input is precisely under our control and not subject to the vagaries of weather or to unknown activities in the lake basin. In a series of experiments where radioactive phosphorus, carbon and iron were added to large enclosures within the lake, we have recently been able to obtain valuable information on how cycles of these critical elements are regulated under natural conditions.

An example is the relationship between iron and phosphorus. Iron is believed to control return of the phosphorus from lake sediments into the water, and hence to be crucial to the management of an eutrophied lake. Again, we were able to correct a long-standing fallacy.

The accepted hypothesis ran as follows. As long as the mudwater interface remains oxygenated, phosphates will form highly insoluble iron salts, so that their concentration in the water will remain low. In lakes which receive large quantities of nutrients, algal growth is stimulated enough to cause large amounts of organic matter to fall to the bottom undecayed. Decaying there, it consumes all the oxygen in the near-bottom waters. Under such conditions, iron in surface sediments is either dissolved or reprecipitated as insoluble ferrous sulphide; either way, any phosphate tied to the iron should be freed into solution in the lakes. Once phosphorus has been introduced in enough quantity to render a lake eutrophic, it returns every year to produce new algal blooms. Preventing further phosphorus inputs is therefore insufficient to reverse eutrophication.

We quickly discovered, in Lake 227, that this pessimistic theory was wrong. An elemental budget accounted for all inputs and outputs and no return of phosphorus from sediments was required to balance the elemental budget. To make sure of our facts we devised further experiments. When we labelled plankton with radioactive phosphorus, none of the phosphorus carried in dead plankton to the bottom ever returned to the overlying water, regardless of whether the sediment surface was oxygenated or not.

Next, we produced characteristic iron-phosphorus salts, containing both radioactive iron and radioactive phosphorus. A SCUBA diver sprayed them as a precipitate directly on the sediment surface in an enclosure in Lake 227. The water overlying sediments was anoxic at the time. At first, much of the phosphorus and iron returned from the sediments to overlying water as predicted by the theory, but both were quickly taken up by particulate matter in the water (probably algae and bacteria) and re-sedimented. Less than 1 per cent of the radioactive phosphorus added to the sediments ever reached the surface zone of the lake. So it turns out that natural chemical control of the phosphorus cycle may be superseded by biological control when the occasion demands. Escaping phosphate is simply taken up rapidly by particulate matter and becomes subject to sedimentation again (Schindler and Lean 1972).

Experiment and management

One of the objectives of our studies with experimental environments is to provide a sound body of evidence upon which resource management can be based – in other words, to fill a role very similar to that played by small experimental farms in agriculture. As the experiments I have described make clear, any attempt to remove carbon from influents of eutrophied lakes would be superfluous; as long as other nutrients are abundant enough a lake can draw ample carbon from the atmosphere to satisfy the demands of its phytoplankton. Likewise, we have disposed of the widespread belief that eutrophied lakes are irrecoverable and that attempts to reduce their phosphorus input are pointless because of the phosphorus 'feedback' from sediments.

Experimental studies of ecosystems on land also help in evaluating current and possible future environmental practices. For example, Herbert Bormann and Gene Likens (1970) investigated the effects of logging a small experimental tract of forest in New Hampshire and of suppressing the underbrush with herbicide. Destruction of the vegetation markedly increased runoff of water from the watershed, as well as chemical concentrations of several ions. Nitrogen in particular reached dangerous levels in streams draining the cleared basins, representing both a loss of valuable soil nutrient and a potential eutrophication hazard to any small lake which might happen to lie downstream.

R.L. Fredriksen (1971) of the United States Forest Service conducted similar experiments, but more specifically designed to simulate commercial logging practices. He found high losses of nitrogen from watersheds on which the residual vegetation had been burned after logging, thus demonstrating that this common forest-management practice was unsound. He recommended that 'slash burning' be avoided whenever possible, because the residual vegetation took up nitrogen and prevented its loss, on unburned watersheds.

Observational and laboratory methods of environmental science have failed to provide a solid body of factual evidence upon which sound schemes for management of natural resources can be based. It may seem unfair, academically speaking, to expect conclusive results from a young science faced with the most overwhelming scientific task ever to confront man. Yet supplies of water, forests and other 'renewable' resources are dwindling at an alarming, ever-accelerating rate, and we cannot afford to wait the tens or hundreds of years which the traditional methods might require in order to provide answers.

Experiments with 'life-size' environments therefore seem indispensable. Meanwhile, the circumstantial evidence provided by the traditional approaches to environmental science is dangerous in other ways. In some cases, such as the supposition that phosphate needs to reach a certain concentration before aquatic plants can use it, the inferences can be seriously misleading. More generally, insecure and ambiguous observations often provide just enough superficial evidence to be manipulated by clever politicians and industrialists into an appearance of environmental concern, while environmental legislation is still subordinated to commercial and political considerations. As Doderer observed: 'Philosophy (science) used to be the handmaiden of theology; now science threatens to become the harlot of politics' (*The Blinding Light*).

References and further reading

F. H. Bormann and G. E. Likens, 'The nutrient cycle of an ecosystem', *Scientific American*, 223 (1970), 10, pp. 92–101.

S.P. Chu, 'The influence of the mineral composition of the medium on the growth of planktonic algae. II. The influence of the concentration of inorganic nitrogen and phosphate phosphorus', *Journal of Ecology*, 31 (1943), pp. 109–48.

R.L. Fredriksen, 'Comparative chemical water quality – natural and disturbed streams following logging and slash burning', *Forest Land Uses and Stream Environment* (Corvallis, Oregon 1971).

G.E. Hutchinson, *A Treatise on Limnology*, 1 (New York and London 1957).

G.E. Likens, *Nutrients and Eutrophication: The Limiting Nutrient Controversy* (Lawrence, Kansas 1972).

D.W. Schindler, F.A.J. Armstrong, S.K. Holmgren and G.J. Brunskill, 'Eutrophication of Lake 227, Experimental Lakes Area, north-western Ontario, by addition of phosphate and nitrate', *Journal of the Fisheries Research Board of Canada*, 28 (1971), pp. 1,763–82.

D.W. Schindler and D.R.S. Lean, 'Biological and chemical mechanisms in the eutrophication of freshwater lakes', *Proceedings of the Colloquium on the Hudson Estuary*, ed. O.A. Roels (New York 1972).

PREDATORS, PARASITES AND POPULATIONS

C. B. Huffaker

Among the more profound of natural phenomena is the tendency of biological systems to attain a high degree of stability, or homeostasis. Within the individual organism, a physiological homeostasis is so essential that life would soon end in its absence and, by natural selection, intricate mechanisms have arisen to adjust for even minor disturbances. Thus, the heart beats faster under greater demand by the body for oxygen.

Here we are interested in homeostasis beyond the individual; that is to say, in the homeostasis of populations and communities of organisms, brought about through more complex natural selection. The fitness achieved is related to a host of interactions among two species or many, associated in natural communities. As Doutt and DeBach (1964) put it:

Every student of biology is aware that in any given locality certain species are more or less consistently abundant, others are less common, and finally, some species are so rarely encountered as to become collectors' items. This condition among the resident species tends to exist year after year, albeit relative and

C. B. Huffaker is president of the International Organization for Biological Control. He is also director of the International Center for Biological Control at Albany, California, an institute of the University of California associated with the Berkeley and Riverside campuses. He is professor of entomology at Berkeley.

absolute numbers vary, but on the average really substantial changes rarely occur in the numerical relationship between the several species inhabiting a given, more or less stable, environment.

We wish to explore the nature of this stability, in well integrated, adapted communities, as well as the consequences of disturbances. Profound perturbations are commonly associated with great and unusual climatic events, invasions by new organisms or alterations of the environment brought about by man. Other disturbances are readily met by natural compensations tending to restore the density of the component species and the character of the community.

Many apparent disturbances are in fact intrinsic to the homeostatic process itself. For example, R.F.Morris and his associates (1963) in eastern Canada present evidence which suggests that outbreaks of spruce budworm may serve such a role. Despite its name, this insect prefers fir to spruce. Even-aged maturing stands, predominantly of fir, suffer more than do mixtures of fir and spruce or stands of fir of mixed ages. Stands also containing hardwoods are particularly less susceptible. Timbering operations have tended to encourage even-aged stands of relatively fewer species. The effect of outbreaks of spruce budworm, then, is to restore a more pristine mixture of conifers and hardwoods, of diverse ages, which is then less prone to outbreaks.

The natural vegetation of the Earth, in fact, presents an enormous potential for compensating for the ravages made from time to time by phytophagous (plant-eating) predators. Mixtures of species are the norm, even though one or two species may predominate in some communities. Plants can regenerate lost parts or new plants can fill gaps in the plant cover, while a vast store of seeds of diverse kinds of plants lies dormant, awaiting any vacancy. In addition, the prey-predator-parasite relation, interpreted broadly, has a truly powerful role in nature. Thus, the vegetation complex results not only from the competition of the plant species for light, water and nutrients, but also from forces acting indirectly, including subtle genetic feedback between each plant species, its natural enemy, and the latter's natural enemies (Huffaker 1971a).

Man not only grows single crop species as vast monocultures but recently he has come to narrow down the crop species to a single variety. Hitherto, a multitude of crop varieties have been selected and used partly because of their resistance to insect pests and diseases. Some six hundred varieties of rice are grown in Indonesia, carrying a correspondingly versatile pool of pest-resistance factors. With the development of the Green Revolution this number may be reduced to only a few, highest-yielding varieties, laying the ground for a potential outbreak of disease or insect pests of an extent formerly impossible. The southern corn-blight epidemic of 1970 in the United States was the most extensive and disastrous single pest-caused catastrophe in the history of agriculture. It was made possible by the extremely narrow genetic base of the 1970 corn crop, about 90 per cent of which carried a common genetic source of susceptibility (Smith 1972).

An individual species of plant is commonly protected by three principal mechanisms. The first line of defence is the inherent, evolved resistance of the plant to the various pathogens, insects and browsers. When those enemies have met and overcome the challenge of the plant's resistance, biological control of the predators by other organisms becomes the chief defence. If that is insufficient, the plant species in question is reduced in density by the predators, until it is sufficiently spaced-out to reduce the pressure of the predators upon it, thus earning inherent protection. In pristine natural communities we would expect that these mechanisms would operate at great efficiency and there are, indeed, few outbreaks of phytophagous insects in tropical rain forests. Changes wrought by man on the environment, as we shall see, can drastically alter this picture.

Natural control of populations

For all organisms, the environment presents definite limits to their numbers. Some species are self-regulating (Wynne-Edwards 1962) and react in behavioural or physiological ways to their own population density in time to avoid the severest environmental penalties. For a great many species, however, control by intense food shortage, by inadequacy of living places, by severe disease, or by predation is characteristic. Food shortage, for example, is certainly a limiting factor for the most efficient invertebrate predators of our plant-feeding insect pests. These mechanisms do not operate independently of changing environmental conditions. Natural control, maintaining relatively stable populations, results from the combined actions of all factors in the environment. Necessarily, though, these factors include at least one component that is density-related and produces a proportionately more intense depressing effect as the population density increases. In short, there must be negative feedback.

Interwoven with the actions of these regulating factors are the properties of the species themselves and various environmental conditions quite unrelated to population density but nevertheless determining the density at which the true regulating factors 'set' population size. For example, in environments where climatic conditions vary widely, populations may spend much of their time suffering and recovering from extremes of the weather and the seasons.

For the rather complex mechanism of control by predators, a number of theories exist, but all have weaknesses. Most have presumed an oversimplified environment and make models of the process that ignore the facts that populations are not homogeneous and do not behave in constant ways, and that the environment is neither homogeneous nor very small and the problems of getting from point A to point B may be vastly different from those of getting from point A to point C. As a result, the capacity of the predator to overexploit its prey is exaggerated and these simplistic models commonly predict self-annihilation of the system. The fox wipes out the rabbits and dies of hunger.

Various mechanisms have been discovered which bring theory into closer relationship with the observed examples of stable, long-enduring control of populations by effective natural enemies (DeBach 1964; Huffaker 1971b). Certain types of behavioural response of a predator to diminishing density of the prey serve to damp the consequences of predation; for instance, the predator may shift its attention to areas of greater prey density. Other damping processes include mutual interference among the searching predators, and also competition for food or shelter on the part of the prey species, in which case the depredation makes life easier for the survivors. Again, if the prey has other, more generalized natural enemies, which tend to switch prey species as their abundances vary, diminution of the population due to one predator may reduce the pressure from these other predators.

Tests of stability

To illustrate what is meant by density-dependent regulation of population size, and by the concept of stable population, we need only alter a population's size drastically, up or down, or change a controlling factor. When we overstock a pheasant farm or trout stream the subsidence of the population to the carrying capacity of the environment is a foregone conclusion. In a natural wood we may find most insects and mites at relatively low and stable densities. If we then apply an insecticide such as DDT, most species will be greatly depressed but this new reduced level does not persist. For most species there is a rapid resurgence to the original levels; for those species whose most effective natural enemies were annihilated, the increase may be to densities much higher than before. Insecticides are commonly used experimentally as a means of proving the disturbing effect of such materials on a stable, non-injurious insect 'pest' situation and of showing the value of biological control by natural enemies.

Recently, in New Zealand, R.T. Paine (1971) has demonstrated the stability of a natural marine community by two manipulations which greatly disturbed that stability. First, he manually removed the carnivorous starfish *Stichaster australis* from a stretch of shore for nine months. Even in this short period, the mussel *Perna canaliculus* increased its distribution by 40 per cent of the available range, and, in this invaded area, eliminated six out of twenty other species. Secondly, from two other areas, Paine removed both the starfish and the characteristic large brown alga *Durvillea antarctica*. Within fifteen months, in both areas, about three-quarters of the available space was occupied by the mussel, to the 'almost total exclusion' of other species. The characteristic distribution and diversity of species in a relatively stable marine community can thus be seen to depend on predation and competition between species.

Predation resulting in biological control of a plant species on land can also markedly influence the composition and species-richness of the flora. This was demonstrated with the introduction into California of the leaf-eating Klamath weed bettle, *Chrysolina quadrigemina*, in 1946. The inadvertent introduction and spread over rangelands by the Klamath weed or St Johnswort (*Hypericum perforatum*) had previously converted a semi-natural stand of nutritious native grasses and naturalized Mediterranean grasses and forbs to a vegetation dominated heavily in places by St Johnswort. Since the beetle reduced the alien weed to negligible (though still common) occurrence, there has been a significant return of the native grasses in some areas and, in others, a re-establishment of the annual grasses and forbs of largely Mediterranean origin (Huffaker 1970). Over a ten-year period there was a significant increase in the species-richness of the range vegetation, towards the condition in comparable range areas that had not become infested with the weed.

Natural control in unnatural situations

Most of the Earth's vegetation now consists of man-disturbed communities. In fact, as M. E. Solomon has aptly remarked, man has prepared 'outbreaks' of brussels sprouts and potatoes and these invite outbreaks of pests upon them. It is indeed hopeful that, in spite of the pervasive spoilage of nature, many intricate natural control devices still operate, even in disturbed systems. True, the natural balance that exists in an alfalfa field or olive grove is by no means the same as that existing in the pristine natural communities where wild alfalfa and olive were indigenous and in which there was natural control of the alfalfa and olive populations as well as of their enemies. Man is fortunate in being able to make use of many of the host-specific enemies that existed in natural communities, and to set them to similar or even better work for him, in his monocultural excesses. We shall now discuss some of the ways in which man has produced perturbed systems, and ways by which he can make some corrections.

The effects of man's relentless drive against all big-game predators are too seldom appreciated. A consequence of his removal of wolves, for example, from most of the United States has been illuminated by recent work in Canada and on Isle Royale in Lake Superior (Pimlott 1967; Mech 1966). In the absence of wolves from Isle Royale, moose became so abundant that they severely overexploited their food resources and the island could not support a healthy sizable herd. Wolves were then established on the island and they have now stabilized the population of moose and their own numbers, and caused a recovery of the vegetation to a level maintaining a healthy and larger herd of moose. The moose population is now rather precisely in balance with the supporting browse potential and the controlling wolf population. The wolf is seen almost as an animal husbander, tending to maximize the yield to himself from the basic plant-food base through the moose, and stabilizing a large complex of lesser organisms. Approximately 22 or 24 wolves now regulate the moose herd at about 625 (in late winter) and take about 125 moose per year as yield, concentrating on the young, diseased and old.

In California's Central Valley, grape growing has caused a tremendous perturbation of the natural control that existed over a formerly unknown insect, the grape leafhopper, *Erythroneura elegantula*. This has been occasioned by the growing of European varieties of grapes as extensive monocultures, with the aid of irrigation, in environments where wild grapes do not occur. The leafhopper is indigenous to Californian wild grape which, in this semi-desert region, grows only near rivers. R.L. Doutt and his associates (1969) have revealed the intriguing story of why this leaf-hopper is a pest in the cultivated, irrigated grapes but causes no damage to wild grapes in their riverine habitats.

In these natural habitats a tiny wasp-like egg parasite, *Anagrus epos*, causes a high mortality of leafhopper eggs, keeping them under excellent biological control. It also achieves good biological control in cultivated vineyards near the rivers. But for overwintering this parasite of the grape leafhopper has to lay its eggs in another leafhopper, *Dikrella cruentata*. This second host occurs, not on wild grapes, but on wild evergreen blackberries growing alongside the wild grapes in the same natural habitats. Doutt and his colleagues have shown that when suitable varieties of blackberries are planted in small patches near the cultivated grapes the parasite, even though far removed from its riverine haunts, can survive and successfully attack leafhopper eggs in the surrounding vineyards. Although this stratagem has not yet been adopted by the industry, its potential value is clear.

Climatic effects in biological control are illustrated by pro-grammes involving crop pests in California. Olive parlatoria scale, caused by an insect native in Eurasia, became established in California in the mid-1930s and soon threatened the olive industry and some two hundred other host plants. No adequate chemical means of its control was then known. In 1951 a number of species of parasites and predators were introduced to control it. Many different stocks of a tiny parasite (*Aphytis maculicornis* or undiffer-entiated sibling species) were obtained from many parts of Eurasia and extensively colonized. Stocks from Spain, Italy, Egypt, India, Pakistan, Greece and other places produced no noticeable control; they were not well adapted to the hot, semi-desert conditions of California. Upon the introduction of the 'Persian' form, from Iran and Iraq, remarkable build-up of parasites and highly effective control were experienced in many olive groves.

Even this highly adapted form is inefficient in hot conditions and it fails effectively to parasitize the spring generation of scales, although it is remarkably effective on the fall generation. Depending upon local conditions, its year-around effect was sometimes marginal or quite inadequate. The solution was to secure a parasite better adapted to summer conditions and thus able to supplement the good work of *A. maculicornis*. *Coccophagoides utilis* from Pakistan is highly adapted to survive and be adequately effective on the spring generation of scales, although less effective than *A. maculicornis* on a year-around basis. The two parasites used together have achieved complete control of this serious olive scale on all host plants where disturbing pesticides are not used (DeBach et al. 1971*b*). The possibilities of finding such pre-adapted or adaptable stocks of parasites open broad horizons in pest control.

The explosive potential inherent to the introduction of new organisms to an environment has been the subject of extensive writings. Charles Elton (1958) reviewed some of the classical examples. When a potentially disturbing organism arrives from its indigenous area unaccompanied by any of its natural enemies the consequences are often grave. They may remain so even when extensive efforts are made to introduce appropriate natural enemies; more commonly, appropriate efforts have been limited.

The inadvertent establishment of the Asiatic fungus *Endothia parasitica* in North America produced the tragic loss of one of the dominant deciduous-forest trees, the American chestnut. The Asiatic chestnut is highly resistant to this fungus and cross-breeding has produced chestnuts combining many of the desirable characteristics of the American chestnut with a high degree of resistance; this does not, however, achieve the re-establishment of chestnut as a self-generative component of the natural deciduous forests.

The deliberate introduction of North American cacti and the European rabbit into Australia resulted in two catastrophes widely differing in character but amenable to similar solutions. Some sixty million acres of agricultural land were ultimately devastated by the *Opuntia* cacti, but the problem was finally solved by introduction of a natural enemy of closely related cacti in Argentina – *Cactoblastis cactorum*. The North American natural enemies of *Opuntia* were not as well adapted to the Australian situation as was *Cactoblastis*.

The story of the rabbit in Australia is the classic example of eventual biological control of a vertebrate pest by introduction of a pathogen, in this case myxomatosis virus. Again, this organism did not evolve in association with the European rabbit but with closely related South American 'rabbits'. While some resistance has developed and some lessening of the virulence of the virus has occurred over the past three decades, the biological control appears still to be highly successful.

This account may serve in a small way to enlighten man of some of the potentials residing in his own nature as a social animal responsive, as are all organisms, to the inherent pressures of his own numbers. Moreover, it is with a touch of hope that we are able to point to ways by which man may correct some of his deliberate abuses as well as the indirect or inadvertent consequences of his existence and his mistakes. It is not without irony, though, that biological control applied in agriculture helps to sustain the human assault on natural systems, whence all methods of biological control derive. Perhaps not too late, man may reduce his urge to take all the Earth's resources as his own and to displace in one way or another any organism that does not obviously 'serve his purposes'. He could start by learning to control his own numbers – using some form of self-regulation involving a conscious, planned programme of which man alone, of all the animals, is capable.

References and further reading

P. DeBach (ed.), *Biological Control of Insect Pests and Weeds* (New York and London 1964).
P. DeBach, D. Rosen and C. E. Kennett, 'Biological control of coccids by introduced natural enemies', *Biological Control*, ed. C. B. Huffaker (New York and London 1971*b*), pp. 165–94.
R. L. Doutt and P. DeBach, 'Some biological control concepts and questions', *Biological Control of Insect Pests and Weeds*, ed. P. DeBach (New York and London 1964).
R. L. Doutt, J. Nakata and F. E. Skinner, 'Integrated pest control in grapes', *California Agriculture*, 23 (4) (1969), pp. 4 and 16.
C. S. Elton, *The Ecology of Invasions by Animals and Plants* (London 1958).
C. B. Huffaker, 'Life against life – nature's pest control scheme', *Environmental Research*, 3 (1970), pp. 162–75.
C. B. Huffaker, 'The phenomenon of predation and its role in nature',

Dynamics of Populations, ed. P.J.den Boer and G.R.Gradwell (Wageningen 1971*a*), pp. 327–43.

C.B.Huffaker (ed.), *Biological Control* (New York and London 1971*b*).

A.MacFadyen, *Animal Ecology: Aims and Methods* (London 1957).

L.D.Mech, *The Wolves of Isle Royale,* United States National Park Service Fauna Series 7 (1966).

R.F.Morris (ed.), 'The dynamics of epidemic spruce budworm populations', *Memoirs Entomological Society of Canada,* 31 (1963).

R.T.Paine, 'A short-term experimental investigation of resource partitioning in a New Zealand rocky intertidal habitat', *Ecology,* 52 (1971), pp. 1096–106.

D.H.Pimlott, 'Wolf predation and ungulate populations', *American Zoologist,* 7 (1967), pp. 267–78.

R.F.Smith, 'The impact of the green revolution on plant protection in tropical and sub-tropical areas', *Bulletin of the Entomologcal Society of America,* 18 (1972), pp. 7–14.

V.C.Wynne-Edwards, *Animal Dispersion in Relation to Social Behaviour* (Edinburgh and London 1962).

HOW WHALES SURVIVE

Ray Gambell

The history of commercial whaling is largely of the successive over-exploitation of one whale stock after another. From the earliest times, whales have been caught more quickly than they can reproduce and grow to replenish the stocks. Current efforts to conserve this natural resource of the high seas illustrate particular scientific and political issues surrounding mankind's use of a shared environment.

Modern whaling fleets are equipped with many navigational and technological aids, including radios for communication between the hunting vessels and for locatng buoyed carcasses, asdic sets for following the whales while they are out of sight underwater, and spotter aircraft to extend the searching area. These, with the fast catcher boats and advanced factory technology now employed, ensure that just about every whale seen can be killed and then efficiently processed for the oil, meat and meal products which find a wide range of applications for industrial and domestic consumers.

The whaling industry is capable of depleting the stocks of whales to the point at which there is a risk that they will become extinct. Fears for the survival of some species are at present fortunately unfounded, but only because long years of negotiations between nations and individual whaling companies are now backed up by adequate scientific assessments of the whale stocks. These data

Ray Gambell is in charge of the Whale Research Unit of the (UK) National Institute of Oceanography and has been responsible for recent discoveries concerning the biology of the sperm whale.

enable all concerned to see what the effect of any particular catching policy will be. There is at last the prospect of rational management regimes being developed and implemented, so that the maximum long-term yields can be harvested – but not before some of the component stocks have recovered from gross over-exploitation in the past.

The history of whaling

Records of European whaling go back at least as far as the twelfth century, when the Basques were hunting the black right whales in the Bay of Biscay and later in the North Atlantic. Whaling from open boats propelled by sail or oars, using hand harpoons to kill the slow-swimming species which floated when dead, depleted the coastal stocks and drove the whalers to more distant waters. The Greenland fishery started in the early 1600s, but ended early in the present century for lack of the Greenland right whales. Similarly, the world-wide fishery for sperm whales, right whales and humpbacks, which started in the late eighteenth century with a large American contribution, also petered out in this century, although the dis-covery of petroleum was a factor in its demise. The Japanese, too, have a long history of whaling and were catching gray and humpback whales off their coasts in the seventeenth century.

Whaling for the fast-swimming blue, fin and sei whales dates from the 1860s, when Svend Foyn invented the explosive harpoon, cannon and steam whale-catcher. Blue whales, the largest animals which have ever lived, formed the first objective of the modern industry. As their numbers dwindled under the impact of the fishery, attention turned increasingly to the fin whales; as these too declined, the smaller sei whales and most recently the minke whales have been hunted more intensively.

As the northern hemisphere resources became depleted, the abundant Antarctic stocks were fished first in 1904 from South Georgia, and then from other land stations and moored factory ships in the Falkland Island Dependencies. Because of the dangers of overfishing, the British government controlled the number of factories operating from its territories and adjacent waters, but these restrictions encouraged the development of the 'pelagic' whaling fleets, which could be free from such regulation. Whaling on the high seas by free-ranging factory ships and attendant catcher vessels began in 1925. The rapid development which followed, from two floating factories in 1925–6 to forty-one factories in 1930–1, led to gross over-production in the latter season, so that the world market for whale oil collapsed. As a result, British and Norwegian companies agreed amongst themselves to limit their production to try to stabilize the industry. In addition, an international convention for the regulation of whaling came into operation in 1935, which gave total protection to right whales, suckling calves and their mothers. Neither Germany nor Japan joined the convention and they were not bound by it or the production agreement when they started whaling in the Antarctic soon after.

Minimum-size limits, a limited whaling season and area were subsequently adopted by international agreements, as well as total protection for the gray whale and some protection for the humpbacks. The blue whale stocks in the Antarctic continued to decline rapidly, and the brief respite afforded by the Second World War allowed little improvement in their position.

The International Whaling Commission

In 1949 the International Whaling Commission (IWC) was set up, as a voluntary association of nations concerned with the whaling industry. It is charged with the task of providing for the conservation, development and optimum utilization of the whale resources. Its regulations are based on scientific evidence, but also take into account the interests of consumers and of the industry itself. Not all whaling countries are members of the IWC, however, and no member nation which objects to a decision need be bound by it. Nevertheless, size regulations and limitations of season and area, as well as protection of gray and right whales, were agreed from the outset.

The IWC established catch quotas in the Antarctic based on the oil production system developed earlier by the British and Norwegians, the blue whale unit (BWU). One BWU equalled 1 blue whale, 2 fin, 2½ humpbacks or 6 sei whales. Unfortunately, this allowed the blue and humpback stocks to be further depleted, because it took no account of the varying degrees of protection needed by each species. The initial quota of 16,000 BWU (some two-thirds of the pre-war catch) was clearly still too high, and the figure was gradually reduced in response to recommendations from the Scientific Committee of the IWC. But the Netherlands refused to believe that the stocks were being overfished and this lack of unanimity resulted in the depletion continuing through the IWC's failure to act boldly enough, soon enough.

Because the IWC can only set a total Antarctic quota, and not one for each nation or factory, there was great competition at first to achieve the largest possible catch during the limited whaling season. This prevented the most effective use of the whale carcases, and attempts were made outside the IWC to agree on national quotas. Initial failure in this endeavour led to the resignation of Norway and the Netherlands from the IWC in 1959. They both rejoined again a little later, and national quotas were finally agreed in 1962. Meanwhile, the whale stocks declined still further and the Scientific Committee continued to recommend reductions in the catches, but without the backing of the Netherlands.

Eventually, in 1960, the IWC decided that there should be a complete re-assessment of all the data available on the state of the stocks of whales in the Antarctic. A special Committee of Three scientists, expert in population dynamics and from countries not engaged in the Antarctic fishery, was appointed to examine the evidence and to report on the levels of sustainable yield and any conservation measures necessary. All the biological and statistical data available around the world were pooled for their analyses, which represented the examination of some 40,000 whales by biologists, and the catch data from 800,000 whales taken over 30 seasons. In response to these assessments the Antarctic quota was reduced to 10,000 BWU in 1963–4, any lower figure being resisted on the grounds that it was economically unacceptable to the whaling industry. In addition, blue whales were given partial and humpback whales total protection.

This still allowed the fin whales to be overtaxed, and in 1965 it was agreed that, along with total protection of the blue whales, the quota should be 4,500 BWU for the following season, with further reductions in the next two years to bring the catch below the combined sustainable yields of the fin- and sei-whale stocks. This aim was reached in 1967–8 with a quota of 3,200 BWU, although the figure has had to be progressively reduced in the light of subsequent scientific advice to 2,300 BWU in 1970–1. The further objective of the scientists, setting quotas by individual species, was at last attained in 1972.

The IWC has stated its intention of bringing all stocks of whales to levels which will provide the maximum long-term sustainable yields. It has taken many years of research and diplomacy to whittle down the catches of a multi-million-pound industry to the point where this aim can be realized, in the face of considerable economic and political pressures. In the meantime the stocks of some species have been very seriously reduced, but present catch limits are now in line with this policy.

The lives of the whales

Anxiety that the large catches taken by the expanding Antarctic whaling operations in the 1920s would result in over-exploitation led to two important scientific developments. In 1924 the British government set up a special scientific body, the Discovery Committee, to study the biology of whales and their environment. The pioneer investigations of these scientists laid the foundations for much of our present knowledge of the whales. Then, in 1929, the Bureau of International Whaling Statistics was established in Norway to collect and publish statistics of catches all over the world. Scientific investigations by several national research groups, together with the statistical data of the catches, are nowadays combined with the mathematical techniques of population dynamics. These techniques, as developed and extended by the Committee of Three in the early 1960s, underpin our current understanding of the whale stocks.

The Antarctic fishery for the various species of baleen whales (the group which includes the blue, fin, sei, humpback and right whales) takes place during the summer months, when the whales are feeding on the dense shoals of shrimp-like krill which swarm in this region. The breeding grounds of these whales are in temperate waters several thousand miles to the north, where the animals migrate for the winter months. Here mating and calving occur but there is relatively little food available and the whales use up the food reserves stored in their bodies, especially the blubber. The female baleen whale carries a single foetus for nearly twelve months, during which time she goes to the polar feeding grounds for the summer, and then returns again in winter to give birth in the warm water of the breeding grounds. She suckles the calf for about six or seven months, in most species, by which time the whales have migrated once more to the Antarctic, where the calf is weaned. The female then has a short pause of five or six months in her breeding activity before mating again in the following winter. The normal rate of reproduction is therefore one calf every two years, in these species, geared to the rhythm of the annual migrations.

Sperm whales differ in many respects from the baleen whales. The female carries the foetus for nearly fifteen months and then suckles the calf for two years, before a resting period completes her normal four-year cycle. Sperm whales do not carry out such wide-ranging annual migrations, perhaps because the squid and fish they eat are more generally abundant. They have a complex social organization, with nursery schools of females and small males which are joined by a harem-master bull during the winter breeding season in the temperate seas. Medium-sized bulls form bachelor schools, also in these warm waters. The biggest bulls are much more solitary and only they are found in the polar regions.

Most of our knowledge of the basic biology of the large whales is derived from observation and collections of anatomical material obtained during the normal commercial whaling operations, since it is not generally practicable for scientists to attempt to carry out such studies elsewhere. This means that there are gaps in the geographical coverage available, and there is no direct, comprehensive information on the stock limits of the various species.

Nevertheless, from studies on the ovaries, mammary glands and reproductive tract of the females, including the size of any foetus present, and of the testes and associated organs of the males, the basic parameters of reproduction can be deduced. Annual growth layers present in the earplug of baleen whales and the teeth of sperm whales give the animals' ages, leading to estimates of mortality rates due to both natural causes and the whale fishery. Knowledge of the migrations of the various species is based on recoveries of special whale marks, which have been found up to thirty years after implantation, together with sightings at sea and abundance records in the catch statistics. From all these data a fairly complete picture has been built up of the life histories of the various species and of key factors needed in estimating the stocks, using theoretical models of the populations.

Stocks and sustainable yields

The two main aims of stock assessments are to find how many whales there are in any given population, and what the effect of catching is on that population.

Estimates of the numbers of animals available in a particular whaling area are obtained in a variety of ways. These include direct sightings by special scouting and survey vessels, and this is the only method by which species protected from catching can be counted. For stocks available to exploitation, the recovery of the whale marks which give information on the migrations of the animals can also be used for estimating the numbers. The most important and reliable estimates, though, are based on mathematical treatment of statistics of catches and fishing effort.

If one season's catch noticeably reduces the population, next season the whales will be harder to find. Changes in the catch per unit of effort from season to season as a result of known catches can thus be used to estimate the original size of the stock. If data for the ages of the catches are available, a number of other techniques can also be employed, including methods that take account of the proportion of the stock removed by a known amount of catching effort, and of the numbers of young recruits joining the exploited stock. All of these ways of estimating the numbers of whales in a stock are 'smudged' by variability in the basic data, although certain factors can be allowed for, such as changes in catching areas, weather conditions and catching efficiency. All the methods feasible for a particular fishery are used to produce a range of estimates. The true value is thought likely to fall within the range and is generally taken as the average of all the estimates when a single answer is required.

Rational management of the whale stocks is now based on the concept of the sustainable yield. In an ideal, unfished population the number of new recruits exactly balances the number of whales dying naturally, to produce a stable situation, limited by food supplies and other natural factors. As the stock is reduced by whaling the rate of recruitment increases, and the natural mortality decreases. The resulting surplus of recruits over animals lost through natural deaths represents a yield which can be harvested indefinitely without causing any change in the total stock size. Paradoxically, a certain depletion of a previously unexploited stock increases the sustainable yield. At some particular stock size, generally about half the unfished populations, the surplus available reaches a peak known as the maximum sustainable yield (MSY).

Smaller sustainable yields are given by stocks both above and below the level corresponding to the MSY, although more than the sustainable yield can be taken from a larger stock, to remove the 'surplus' and so generate the maximum yield. But if more than the sustainable yield is taken from a stock below the MSY stock level, the stock is further reduced in size and becomes even less productive in the long term. At very low stock sizes the recruitment and natural mortality rates become about equal again, so a population which is grossly depleted may not increase in numbers to any significant extent when hunting ceases.

The reduction in natural mortality when a previously unfished population is exploited is probably the result of catching some whales before they have a chance to die naturally. To this may be added better feeding; the reduced stock still has access to the same food supply, which helps the whales to withstand some of the natural causes of death and is probably also an important factor for raising the recruitment rate in exploited stocks. The whales grow faster and become sexually mature at an earlier age; the adults also become pregnant more frequently. By quantifying these various changes and putting them into the mathematical models of the whale stocks, the sustainable yields at each size and the MSY for each stock can be calculated. The effects of different rates of catching, under varying management schemes, can then be predicted.

Whaling today

The major whaling grounds fished today are the Antarctic and the North Pacific. There is a long history of shore-based whaling in the North Pacific, but it was only with the increase of factory-ship operations by Japan and the Soviet Union that catches rose sharply in this area as the Antarctic situation deteriorated in the 1950s. Catches were regulated in the North Pacific, first by inter-governmental agreements, but now through the IWC. Smaller-scale fisheries are also carried on in all the other oceans of the world, mainly from shore stations.

Throughout the world right whales had been severely reduced, even before the advent of modern whaling. They are now protected from hunting, as are the gray, blue and humpback whales which were greatly depleted, in all the areas in which they occur, in the first half of this century. Apart from the gray whales in the eastern North Pacific, which have increased moderately under protection since the Second World War, the various stocks of all these species will require up to fifty years or more to rebuild their numbers before they can be safely exposed to commercial catching again.

Of the large whales currently being exploited, the fin-whale stocks in the Antarctic and North Pacific are well below the MSY stock level and need to be allowed to rebuild. The fastest recovery will be achieved by catching none at all for a time, perhaps for twenty-five years. Since the fin whales still represent a major part of the whaling industry's catch, such a move will have important economic repercussions. Sei-whale and sperm-whale stocks throughout the world are generally at or above the MSY stock levels, so properly regulated catches of these species can continue.

Each whaling operation has an inspector appointed to it by the government concerned to see that the international regulations are obeyed. An international exchange of observers between the active whaling nations in the IWC also takes place now, to satisfy any doubts about fair play. It is only in this way, when the catching power of the industry is so great, that conservation, rebuilding and rational management of the stocks of the large whales can be achieved on a global and international basis. It requires that all the whaling nations co-operate and refrain from pursuing independent policies. With the past grim record as a reminder, and improved scientific advice as a guide, there is hope that the whale resources of the world will be properly maintained and utilized from now on.

Compared with other problems of managing the environment and its resources, whaling policy is conceptually fairly straightforward. The perils of overfishing have been obvious to scientists for at least half a century, but the powerful national and commercial interests represented in the industry have been hard to convince, to put it mildly, and during the 1950s and 1960s they engaged in a kind of ecological brinkmanship which threatened the downfall of the industry itself. Economically, it may be more profitable to catch nearly all the whales there are in a few seasons, and then close down, rather than operate at a low level for a long period of years. However, the industry's increasing difficulties in finding whales of some species has helped to bring about the recent change of heart, but there is no doubt either that the scientific methods of reckoning whale stocks and sustainable yields had to reach a high level of sophistication and precision before the sceptical and reckless could no longer ignore them. So a general lesson for environmental science is that for confronting vested interests only the best research will do.

References and further reading

International Whaling Commission, *Annual Reports*, 1–22 (London 1950–72).
International Whaling Statistics, 1–68 (Oslo 1930–71).
N. A. Mackintosh, *The Stocks of Whales* (London 1965).
L. H. Matthews (ed.), *The Whale* (London 1968).

MAKING GRASSLAND COMPUTABLE

Norman R. French

All organisms live in groups, in association with others of the same kind and of different kinds. The individual organisms interact with one another in a complex and sometimes subtle fashion. In this way the materials and energy of the natural environment are apportioned among them, and cycled through various levels of the living community. Conditions of the physical environment also influence the organisms, modifying their activity or placing limits upon them. Through their sensory abilities, living organisms obtain cues from the environment and modify their activity for their own protection and benefit. These interactions and feedbacks are so complex that they nearly defy human understanding; yet we rely on them for our survival.

One way of picking at this knotty problem was first set forth by Raymond Lindeman (1942). He introduced the trophic-dynamic concept of ecology. According to this concept, materials and energy are continually moving between different compartments, the different nutritional or 'trophic' levels of the ecosystem. Energy is originally fixed in the system by the green plants, which represent the first trophic level. This energy is then available to other organisms

Norman R. French is professor of biology in the Natural Resource Ecology Laboratory, Colorado State University. He is also director of field studies for the Grassland Biome Program, under the International Biological Programme, which is supported by National Science Foundation Grant GB-31862X.

in the ecosystem, such as the 'consumers', animals that feed upon the plants. These, therefore, represent a second trophic level. Finally, materials are processed by the decomposer trophic level and again made available as nutrients and minerals for recycling through the system. A natural ecosystem contains many other levels and possible channels for energy and nutrients, and organisms regulate and modify these flows on the basis of environmental cues.

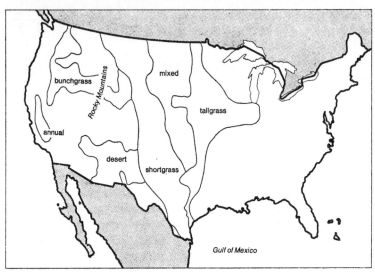

Types of grassland in the United States

Traditionally, biological science has progressed by isolating and focussing attention upon one portion of one compartment of the ecosystem, or upon one segment of one pathway whereby material is transferred through the system. Simplifying the problem in this way has revealed much about how individual organisms operate or react, and the processes that regulate particular flows through important channels of the system. However, the very fact that such studies were conducted in isolation makes it difficult or impossible to use very much of this information in evaluating the many interactions and feedbacks that occur under natural conditions.

An example of the complex interactions between the physical variables, which act as driving forces of the biological world, is the manifestation of climate that determines the distribution of grasslands (see map).

The climate is itself a complex pattern of interacting physical variables. That grasslands occur as a broad band in the middle part of the North American continent is determined basically by the distribution of solar energy and of rainfall, and by the flow of air. While storm tracks out of the west and north-west bring in winter precipitation, the important moisture during the growing season moves northwards from the Gulf of Mexico. This summer moisture is counteracted by a prevalent flow of warm, dry air that has been stripped of its moisture in passage over the mountain barriers to the west. This dry air is manifested in the tongue of grassland protruding eastwards in the central United States. Solar radiation in mid-summer is about 20 per cent greater near the western mountains than at the eastern boundary of the prairie, adding to the drying power of the air and causing an increasing moisture deficit towards the west.

Eastwards, in the tallgrass prairie, there is enough rainfall in the growing season to make good the losses of water; towards the west, however, in the region of mixed and shortgrass prairies, this is not so. The line separating the region of water surplus from the region of water deficiency is essentially the transition zone between the tallgrass and the shortgrass prairie. These conditions result in alternating dry and wet soil and provide the environment in which grasses flourish.

Grasses are relatively new in the history of the Earth, having become a prominent feature of our landscape only in the past twenty million years or so. Their advent caused evolutionary reverberations throughout the consumer populations, and the organisms that quickly adapted to the new conditions had new horizons opened to them. A marvellous variety of grazing herbivores developed.

In North America, for example, a primitive horse named *Merychippus* developed the physique and physiology needed for feeding on grass. This adaptation permitted a descendant group, *Hipparion*, to conquer nearly the whole world by wandering from North America through Asia and even attaining the continent of Africa, before being finally extinguished by unknown misadventures. This conquest was exceeded only by a later descendant, the modern genus of horses, *Equus*, which moved out from North America not only to Asia and other parts of the Old World, but also to the South American continent, recently made accessible by a land bridge. Our present-day horses and zebras are remnants of this world-wide invasion.

A great assemblage of huge herbivores dominated all of the major continents until about a million years ago, when a comparatively small omnivore, without impressive evolutionary credentials, became fairly numerous on the scene. That animal was man. He hunted the native fauna and eventually he developed agriculture and domesticated the animals. He prospered and seemed able to exploit almost any type of environment.

In the early part of the twentieth century, man brought agricultural methods developed in eastern North America to the vast prairies of the mid-continent. There he was defeated. Never lacking in graphic words, he called his failure the Dust Bowl. He then began to learn there were limits to the degree of modification he could impose on the environment, and that natural ecosystems have sometimes developed in an acutely delicate balance with the physical conditions controlling them. He learned quickly, however, and began to develop a scheme of management for the semi-arid plains, or steppe regions, so that he could derive sustenance and benefit from the system without disturbing the balance to his own detriment. The grasslands of western North America are one habitat that has improved under intelligent management practices during the past few decades and which give hope for the future in these days of generally deteriorating environments.

Fostering interdisciplinary research

In an attempt to understand the intricate web of interactions within natural systems, considerable scientific effort is being devoted to modern interdisciplinary studies. These employ team research and computer technology in an attempt to find optimum solutions to environmental problems. The Grassland Biome Program is one such venture and is part of the United States contribution to the International Biological Programme. The term 'biome' designates a broad formation of vegetation, characterized by species of plants with certain common characteristics that predominate over the entire area. In this particular case, grasses are the predominant species and make this biome readily distinguishable from those dominated by deciduous trees or coniferous trees. Considering that there are fourteen million square kilometres of grassland in the world, and that this area fixes one-tenth of all the carbon transferred into organic matter annually, this ecosystem and its relation to man is plainly worthy of systems-analysis and computer modelling. Only the tropical forests and the oceans of the world produce more organic matter than do the grasslands (Lieth, in the press).

The key to success in such a programme is the establishment of communication among the different disciplines represented by the scientists involved. Each scientist needs to consider what impact the organisms he investigates have on other kinds of organisms in the system, and conversely the impact of other organisms on the things dealt with in his own special field. In addition, there is a set of factors which override and drive the whole system and the effects of these on particular organisms must be evaluated.

Consider, for example, the decomposition of organic material, which is essential for the recycling of nutrients. Organic materials produced by the plants may end up in the stems, leaves, or the roots. This ultimate location may have considerable bearing on the rate of return of these materials to the soil. Further, they may be cycled through the consumer population before being returned to the soil. In any case, all of these functions are affected by the temperature and perhaps the water conditions under which they occur, be they growth, decomposition, or consumption by animal populations. So we begin to see that a natural system is actually a network of alternative pathways through which matter and energy may move through this complex – without regard to the prejudices of specialist scientists! (See diagram overleaf.)

Interdisciplinary awareness has provided us with new insight into the dynamics of root material in grasslands (Bartos and Jameson, in the press). Through the growing season there is first a decrease, then an increase in the amount of root material in the first twenty centimetres or so below the surface. Earlier workers, trained to think in terms of individual plants, reasoned that an early-season drain of carbohydrates from the roots helped to produce growth above ground; this, they supposed, was followed later by a reverse flow of material to storage areas in the roots. A new hypothesis has been formulated by scientists taking into consideration not only the mechanisms of plant growth but also the microbial processes of decay. They now believe that old root material undergoes rapid decomposition as soon as soil moisture and temperature conditions become favourable in the spring, causing the initial decrease in root material. Then, as the rate of plant growth increases, new root growth brings about a rapid increase in plant material below ground.

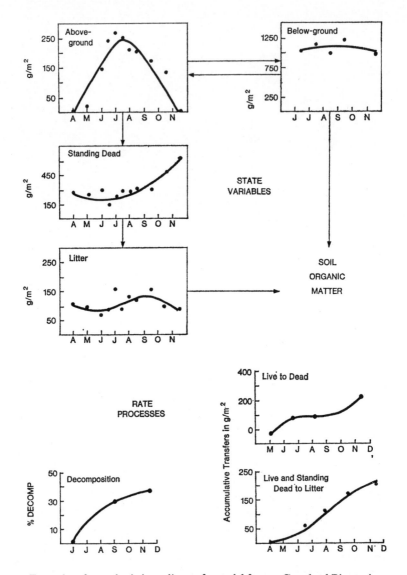

Examples of some basic ingredients of a model for one Grassland Biome site

The monthly changes of some state variables for primary producers are shown, and rates of change through time for transfer of materials from one compartment to another

This hypothesis would probably not have been formulated by plant scientists, had they not learned of the discoveries made by microbiologists, about the factors influencing rates of decomposition of organic materials in soils.

Yet accomplishments in science have traditionally flowed from a high degree of specialization and from the experimental method that simplifies the system and isolates the governing factors. This tradition is reinforced by the enormous amount of highly specialized literature that is generated annually. Positive action is therefore needed to break down the mental walls that divide scientists.

In the US/IBP Grassland Biome Program, we developed interdisciplinary communication through a series of workshop-seminars; these were held in remote locations to encourage informality and much individual contact among the participants. At each seminar, several review papers in one general area represented quite different disciplines. At first, scientists were reluctant to make inquiries of specialists from other fields. A human tendency is to avoid raising questions about subjects with which we have little familiarity, for fear of revealing our ignorance. Once the participants overcame these inhibitions, however, the ideas began to flow quite freely. It was highly stimulating, not unlike the 'mind-expanding' experiences being sought by young people today!

Investigators representing many scientific disciplines have been brought together in a functioning, co-ordinated research endeavour that is uncovering information about many aspects of the grassland ecosystem. In the Grassland Biome Program more than ninety scientists are taking part; they represent the co-operation of thirty or more different institutions and agencies, which marks a new era in the biological sciences. All data and summaries, accumulated in a single laboratory, are available to all participants. Through special efforts at synthesis of these results, comparisons are made among sets of data to arrive at generalizations. We seek not only descriptive information about the details of the system's operation, but also, and more importantly, principles that apply to all grassland systems. These generalizations and principles will form the raw data on which model-building can proceed.

Problems of communication among the group of participating scientists increase at an exponential rate as the number of participants increases. It seems evident that there is some optimum number

where the maximum scientific benefit is derived from each participating scientist. At least one meeting annually must involve all programme participants, and result in a review of the overall activities.
Other meetings with specialist groups must be arranged for searching into details of problems in methodology. Similar specialist
groups must consider the synthesis of information. Those of us who
manage the programme must be involved in all of these activities
to some degree, as well as in planning and strategy sessions.
Efficiency depends on our methods of organization and
communication.

Our pooling of data and results runs counter to the tradition
whereby the scientist conducts a study independently and accumu-
lates and analyses all his data before reporting them to his colleagues.
This is a notoriously inefficient process, geared to the needs of our
academic institutions in evaluating candidates for degrees and
passing judgement on their scientific accomplishments.

We have learnt to maintain a high degree of flexibility. As advances are made and relationships become understood, old problems are solved and new ones are encountered. The priorities of the
programme therefore change continually. This not only means that
emphasis must shift, but also that entire sets of successful experimental plans must be terminated and hopeful new ones instituted.

Detailed research is still at the heart of our programme. The
essential information about the system must be obtained by the
specialist with intimate knowledge of the individual components
of the network. Only he can plan quantitative experiments to obtain
reliable data on the various processes. Sometimes highly sophisticated studies are involved. For years, range managers have known
that certain plants flourish early in the growing season, while others
are slower to mature and reach the peak of their growth later. Recent
investigations of the photosynthetic process in range plants suggest
that these different growth patterns are related to different biochemical routes for carbon fixation, the so-called C_3 and C_4 pathways (Williams 1971). The timing of development of the plant
turns out to be a reliable indicator of the biochemical processes
within it.

Finally, under our programme, the results of efforts in the field
and the laboratory must be pulled together. Functional relationships have to be expressed, in a form appropriate for the models of
the system – often as a graph showing how the rate of a process

depends upon another condition or variable. For example, the rate of decomposition of organic matter is governed by the temperature and water-content of the soil. If these values are known, the rate can be predicted. In the tallgrass prairie, where moisture is ample during the growing season, decomposition of organic material is nearly complete within sixty days. In the shortgrass prairie, with appropriate temperature conditions but limited soil water, decomposition may proceed in spurts, after rainfall. With plenty of water, decomposition proceeds at the same rate as in the tallgrass prairie, but that condition is seldom met in the shortgrass prairie, which therefore acts as a pulsed system.

Such ingredients of knowledge can be built into a model of the system, and these processes can be simulated numerically when the appropriate driving factors fall within a certain range. The model, of course, must eventually be validated against the actual observations made in the field. Only by this cycle of insight, evaluation, experimentation, integration, modelling and eventual testing of the model will we be able to develop management schemes which are suitable for complex situations represented by natural systems.

Some discoveries along the way

An unexpected effect of grazing on the grasslands involved the interrelationship between plant cover, a bird, and grasshoppers. Lark buntings like to nest in the cover of the saltbush and red threeawn grass. During the breeding season each bird will consume as many as 65 grasshoppers, 103 weevils, 122 ants, 33 scarab beetles, and 352 other insects, as well as seeds. Therefore, when cattle reduce the cover and the birds fail to nest, the overgrazing syndrome is worsened by the loss of natural control over the grazing insects. In such subtle ways, interference may reverberate and intensify through various levels of the grassland ecosystem.

A synoptic analysis of the climatology of the Great Plains by Grassland Biome meteorologists has indicated long-term trends superimposed on the ordinary fluctuations in climate from year to year. An unexpected trend is the decrease in solar radiation over this region. Total incoming solar radiation has, on the average, declined by about 10 per cent during the past two decades, possibly due to man-made increases in the density of particles in the atmosphere. Although a change of this magnitude may be inconsequential in terms of the radiation required for the process of photosynthesis and plant growth, recent evidence suggests that species of the shortgrass plains are not, in fact, saturated by light at ordinary intensities. In any case, the decrease in radiation may well have an indirect effect on plant functions as it changes the energy balance near the ground. Most of the radiation is either converted to heat at the surface or else promotes the vaporization of water. Changing this heat balance could have significant effects on the soil water available for plant growth.

What are the implications of such long-term trends? Are the climatic changes due to natural fluctuations or are we truly altering the climate of our planet? Are climatic modifications on the whole detrimental or beneficial? The decline in the deer herds of northern Minnesota is blamed on increasing severity of the winters there. On the other hand, crop failures in the United States have become fewer during the past two decades. In 1956, when droughts reached a peak, nine million hectares of agricultural crops was lost. Since 1968 the figure has been about three million hectares annually. One hypothesis states that a drop of $3.5°C$ in the average annual temperature will trigger another ice age. Ecosystem models which accurately depict year-to-year variations in productivity will help to answer such questions.

For a final example, scientists have recently estimated that a tiny scale insect, the mealybug, subsisting on the sap of prominent grass species, may be just as important a consumer on the range as are cattle. Although each insect ingests only a gram of sap each day, their numbers suggest that this is an important removal of energy. We do not know what the shortgrass prairie would be like without this little consumer. It is possible that net productivity would be significantly altered, and possibly vegetation composition as well. The full intensity of grazing pressure by the rabbit in Britain was revealed only when its numbers fell catastrophically, as a result of infection with myxomatosis.

Projections

The trend of the future seems fairly clear. Large-scale complex problems, such as those involving modification of the environment by man and his technological developments, must be formulated in a systems-analysis framework if they are to be understood sufficiently well to allow wise decisions about the management of natural resources. The fact that alternative strategies will be considered gives rise to the more difficult question of weighing costs and benefits in the different strategies. Although it may be possible to assign an economic value to the development of a power plant or utilization of grazing land, how can a comparative value be estimated for parkland, for example, which may have high aesthetic but low economic value? The evaluation of alternative strategies in multiple-use planning thereby becomes a 'multi-interdisciplinary' problem, involving social scientists as well as specialists who understand the utility of specific resources.

Systems models are instructive as to the nature and complexity of modifications or perturbations to the system. A complex model must, however, be considered only as a working hypothesis. Results from a model may be suggestive, but verification will inevitably await the outcome of field experiments. The reason for this is that it is feasible to model accurately only small portions of systems or subsystems. Only with a rather narrow and limited problem can all the complex interactions be incorporated in a model. Even these must be suspect until rigorous experimental proof has demonstrated their validity under all conditions. A model of a larger system can only mimic the system, at best. By 'mimic' I mean that the principal variables of the system, incorporated into the model, will respond to changes in other variables in a manner similar to that observed in nature.

So we must not be over-ambitious in our model-making. The most serious and intensive research effort might never succeed in understanding the biological meaning of all the interactions in a small patch of grassland. Even if they could be understood, it is questionable that any model could ever be developed that would incorporate all interactions simultaneously. And even if such a model were available, no existing agency could support the cost of running it on a modern computer! Therefore, our models are destined to be highly simplified versions incorporating only the major and most conspicuous variables and interactions of the natural systems. They will nonetheless be of great value in assisting the decision-making process, which will of course go on with or without models, and also in guiding scientific research towards fruitful areas, thereby making better use of scarce resources.

References and further reading

Dale Bartos and Donald Jameson, 'A dynamic root model', *Journal of Range Management* (in the press).

Helmut Lieth, 'Net primary productivity of the earth with special emphasis on the land areas', *Perspectives on primary productivity of the earth*, symposium volume of the American Institute of Biological Sciences (in the press).

Raymond Lindeman, 'The trophic-dynamic aspect of ecology', *Ecology* 23 (1942), p. 399.

George Williams, 'Producer function on the intensive and comprehensive sites', *Preliminary Analysis of Structure and Function in Grasslands*, Range Science Series 10 (Fort Collins, Colorado 1971).

MODELS OF THE HUMAN ECOSYSTEM

Kenneth Watt

The human environment is an interacting system of population, land use, transportation, economic activity and capital allocation, use of energy and other resources, pollution, the effects of pollution on weather, agricultural production, and public health. In this brief essay, the term environmental systems models will be used to describe those models or groups of models that set out to mimic this system in a comprehensive way. Models which attempt this degree of complexity have been under development only since about 1968, and the first major reports on this type of activity appeared in 1971 and 1972.

By a 'model' we mean, of course, not a physical structure but an intricate set of data and mathematical relationships between data which not only describe the present situation but also 'work' in the sense of predicting the consequences of change. The development of this new field has been made possible by great advances in other academic disciplines.

Computer specialists and general systems analysts have provided environmental systems modelling with its methodological tools. The hardware is the new generation of computers with very fast computing speeds, and gigantic memories. The software

Kenneth Watt is professor of zoology in the University of California at Davis. Research by his environmental systems modelling group is supported by National Science Foundation Grant GI-27.

includes the computer languages which make possible the rapid coding of enormously lengthy and complex sequences of instructions: FORTRAN, ALGOL, and the special simulation languages DYNAMO, CSMP, SIMSCRIPT and SIMULA. There are also a host of methodologies from applied mathematics and statistics. Particularly important has been the bringing together of many of these techniques into a powerful group of methods for exploring the way a complex system changes through time in response to policy decisions. Perhaps more than any other individual, the mathematician Richard Bellman was instrumental in demonstrating that computers were useful for exploring the consequences of sequences of decisions.

The whole group of new statistical, mathematical and computer tools is usually referred to as systems analysis. A central idea is that the state of a system at a particular instant in time can be predicted from the state of the system at a previous time, if we can discover a set of equations that realistically simulates the way in which the system changes during the interval.

A second group of advances which led the way into new, comprehensive systems models came from regional planning. This discipline had, by about 1968, developed complex, detailed and realistic models for a number of regions, relating population, land use, transportation and regional economic activity. Other contributions came from ecology and resource management which, from 1963 on, were developing increasingly complex and realistic models for the management of forests, fisheries and grasslands. New data and computer simulation studies in the earth sciences have been important in indicating that pollution could become serious enough to influence climate on a global scale, thereby affecting crop production in any region. Thus a completely realistic regional or global systems model must mimic the production of pollutants by industry and vehicles, the effect of this pollution on climate, and the effect of climate on plant growth. Finally, great advances in economic research in the last few years have led to the development of computer simulation models of national economic activity.

The various fields just mentioned not only provided techniques and insight necessary for building computer models of the human ecosystem; they were also important as the sources of the staff who now, in several institutions, work together in teams of up to fifty people building the new, large models.

One other prerequisite makes possible the new models: the overwhelming masses of data that are routinely collected on almost everything characterizing the modern technological society. The partial models developed by statistical analysis of these data provide the building blocks for the larger systems models that seek to describe a region, or the whole globe, in a comprehensive or holistic manner.

Why holistic models are needed

Realistic holistic systems models depend on an immense amount of statistical data analysis, followed by a great deal of simulation – that is to say, game-playing to explore the consequences of different policies. All of this stretches the capabilities of the largest new computers to their limits. Also, it now appears that upwards of fifty man-years of effort by very well trained people are required to complete such a project. Thus, the total bill for such an endeavour is about 1.5 million dollars. Given this cost, what are the returns? Why couldn't the same benefits be obtained with a set of smaller models, one each for land use, energy production, pollution, economic activity, transportation, and each of the other subcomponents of the entire system? This is a particularly important question, because the answer gives an intuitive feeling for the meaning of 'system'.

Suppose that, within a large group of countries, there is a great increase in energy consumption per capita. This increased energy consumption leads to the generation of increased pollution, from which the net effect is a depression of the global mean temperature. Suppose further that the net effect of this climate change is a general lowering of mean crop yield per hectare. That, in turn, means that more energy per hectare (in the form of fertilizer, pesticides and use of farm machinery) is needed to keep crop yields increasing so that the population can be fed. Thus, greater use of energy per capita ultimately leads to the need for still greater energy consumption per capita: a *positive feedback loop* in which use stimulates more use. However, the closing of the loop can only be perceived if we have a conceptual model which links energy consumption, pollution generation, geophysical processes, weather regimes, crop growth, and technological inputs to agriculture. The actual magnitude of the effects as this wave of causes and effects is propagated through the system can be estimated only with a holistic systems model which incorporates all the components of the system.

Similarly suppose, as in the United States at present because of high birth rates between 1952 and 1964, that there is a very large number of people under twenty years of age in the population, compared with the number of older people. This creates a demand for new jobs that can not be met by the ability of the existing labour force to generate capital; the result is high unemployment amongst young people in the years of life normally associated with very high birth rates (twenty-one to twenty-four years of age). The result is a sharp reduction in birth rates among these young people – there is *negative feedback* acting on the birth rates. While the first effects are experienced by baby-food and diaper manufacturers, subsequent effects are felt in reduced demands for housing, urban land, energy, transportation and agricultural produce. Some other interesting loops are created. Because there are fewer births now, five and more years hence there will be lowered demand for school teachers. If the birth rates persist for more than five years, because of continuing unemployment amongst the young, this means lowered demand for services to the very young, and therefore still more unemployment in the twenty-one to twenty-four age bracket, and still lower birth rates. This feedback is significant, because about 4.3 per cent of all jobs held by women in the United States at present are teaching jobs. A weakness in almost all regional and urban models at present is that they fail to include loops relating the age structure of the population to economic activity.

Some readers will wonder why a large number of unemployed young people doesn't lead to increased demand for goods and services, the provision of which provides jobs sufficient to hire all of them. The reason for the failure of jobs to keep pace with population is the increased concern with efficiency in all business, industry, government, education, transportation and farming organizations. Because of this efficiency, often achieved with automation, increased production of goods and services can be attained without a corresponding increase in staff. The only way to beat this problem is to have a correspondingly rapid shift in deployment of the labour force away from labour extensive employment towards labour intensive service industries. Only through this change in the 'mix' of the economy can rising unemployment be averted.

Holistic systems models also relate age distribution, projections of population growth, demand for energy per capita and known reserves of fossil fuels, to arrive at projections of demand for energy, country by country, for all countries. For each country, population size and demand for energy per capita are projected separately and then multiplied together to obtain total national demand. This is done for each category of energy separately (coal, oil, gas, hydro-electric and nuclear power). The computer can sum demand for each type of energy for all countries in each year, and compare this with the total amount of fuel known to be available in each category. In holistic models, these energy projections can be used to make estimates of the time at which different energy sources will be totally depleted. It should be noted that even nuclear power could run out in a few decades, unless there is rapid conversion to breeder and fusion reactors.

Since 'hidden' energy subsidies to agriculture – use of farm machinery, fertilizer and pesticides – are critically necessary for increased yields, any model of the availability of food will produce unrealistic future projections unless it is linked to an energy supply-demand model, which in turn must be linked to a demographic model. The energy available for agriculture determines how many people the world can support, so no model for projecting future populations is realistic unless it includes the loops which mimic the effects of energy availability on agricultural production. Also, estimates of future demands on resources are unrealistic unless they are based on projections of the way in which population age structure will affect birth rates.

The important common feature in all these cause-effect pathways is that, in each case, the ultimate explanation for a phenomenon or a process is in a different mechanism than that under study. Putting it another way, all parts of a complex system are affected by all other parts, because of a system of reciprocal and circular feedback loops which link up the whole system. Consequently, a cause operating at one point in the system may have types of effects elsewhere which would not be expected unless one had done simulations with a systems model.

Special features of environmental systems models

Several features of this type of research activity will already be apparent. It has a strongly interdisciplinary character, is done by large teams, is highly complicated, and is absolutely dependent on large computers. This type of research also puts strong emphasis on policy – what happens if we pursue policy X instead of policy Y? It puts a great deal of emphasis on the relationships between various components of a system, whereas most scientific research concentrates on particular components. That is to say, in environmental systems modelling, we are not satisfied, for example, to know how changes in age-specific birth rates affect future population projections; we also want to know the implications of these changes for demand for power, housing, transportation and land, and the implications of those changes, in turn, for remaining stocks of fuel, pollution, the weather and crop production.

Systems modelling puts primary emphasis on the structure and dynamic properties of the whole system. We ask ourselves such questions as the following. Which are the driving mechanisms in the system – for example, which has the greatest effect on total agricultural production: availability of land, or availability of energy, and does the answer change for different total population sizes? How closely are different systems components coupled? Are pollution levels having a sufficiently important impact on death rates to justify inclusion of a loop between energy consumption and population size operating through the death rate, as well as one operating through the birth rate?

Because systems modelling deals with policy, it comes very close to politics and the problem arises of communicating the results to a mass lay audience, so that intelligent voting patterns may result. Using present technologies, it is very difficult to communicate our material to the public, because of the volume of the output and also the complicated relationships between different variables. For informing key individuals, however, it is possible to link the computer to a television screen that depicts the output in pictorial rather than tabular fashion. A person – for example, a political scientist or a trainee planner – can sit at such a screen holding a pen-shaped pointer called a light pencil. The computer can then display the state of the situation at a particular point in simulated time and ask which of several different policies should be used to produce optimal changes in the situation as just depicted. This question can be posed by the computer in the form of a list of, say, eight options presented in sentence form on the screen. The trainee indicates his choice by touching the light pencil to the appropriate option on the screen. The computer then responds by simulating the consequences of having chosen this policy. Clearly, this is a

training tool with a tremendous potential for compressing experience into a short time – assuming of course that the simulation
model is realistic – and allowing people to discover their mistakes
from the computer rather than from public discontent.

The reason why this teaching tool is so important is that highly
complex systems have a 'counterintuitive' feature, as Jay W.
Forrester (1971) has argued. Because of the multiplicity of circular
feedback loops found in these big systems (a city, a region, or the
world), the system may often react to a policy change in a wholly
unexpected fashion, because we inadvertently affect other loops
besides the ones we had intended to affect.

A further characteristic of environmental systems modelling is a
great concern with ingenious, highly aggregated *indices* or measures
that reveal the essential properties of the system without including
so much detail that the computer memory is overloaded and the
human mind boggles at the sheer volume of output. Thus, we are
led to ask the question: 'Which are the most critically revealing
measures of events and processes in society?' Two examples will
illustrate the kinds of measures that are most useful.

Knowing how many vehicles pass per hour down each of thirty-
seven different main arteries at peak daily rush hours may be
important for the traffic police but not for the environmental
systems modeller. Rather, it is more revealing to ask what is the
annual energy-cost per person of transportation systems in the
metropolitan region in question. This quantity is the product of
three others multiplied together: the number of passenger trips
per person per year, the average number of miles per passenger trip
and the fuel consumption per passenger-mile. The first two of these
three multipliers express the effect of urban sprawl, and the distribution of work places and homes in the metropolitan area. The third
expresses the effect of choice of car, bus or train and also the impact
of traffic congestion, with frequent starts and stops, on fuel
efficiency.

Nor is it too revealing to know how many hectares of land were
converted each year from one category of land use to another, for
each of twenty-nine urban areas. More significant is the conversion
rates *per person* per year, for each of the areas. Then one can express
the impact of various other factors on these rates and bring out very
clearly the impact of various zoning and land-use policies on the
efficiency of land use; a further step links this with a transportation

analysis for comparisons of the efficiency of energy use in transportation.

Sensitivity analysis is yet another special feature of environmental systems modelling. We ask which of a number of variables has the largest impact on another variable. For example, how can we best make more efficient use of energy, by switching from cars to trains, or by cutting out the use of plastics, or cutting out the use of high-tensile-strength alloys? We also ask how much effect a particular trend, or policy decision, or error in forecasting, may have on the system. For example, it may make rather little difference to certain projections if the world's population takes a decade longer to double than current estimates suggest; again, the availability of public transport is highly *decision-sensitive*, while the energy per passenger-mile for a bus is *decision-insensitive*.

What can modelling accomplish?

Many readers may be asking themselves: 'Of what use can this type of research be, when it relies on information from other areas of research which are themselves full of unresolved controversies?' For example, demographers cannot agree as to whether the population of a particular country or region will increase rapidly or slowly in the future, or decline. Geophysicists disagree as to whether man-made particles in the air will cool the planet more than man-made carbon dioxide will heat it up. Economists disagree as to whether the economy will improve or get worse. Agricultural experts disagree as to the rate at which prime farmland is being converted to city use. Environmental systems modelling cannot resolve this type of controversy. However it can find the possible range of impacts of one subsystem on another, so that if the high estimates of a process are right, we know the implications for other processes, but if the low estimates are right, we know the implications of those also. This is merely another way of saying that the relationship between systems subcomponents are of more interest to the modeller than the values of the systems subcomponents themselves.

Systems modelling cannot predict the future. Different futures are possible, each following from different policy decisions. These decisions might be made by legislative, executive or regulatory bodies, or by the electorate, or by the public without operating through government; an example of the last type is the decision to have fewer children, or to boycott producers of grapes or the manufacturers of certain makes of cars. With models we can explore the implications of such choices.

Three examples illustrate the type of results that can be expected from this class of research. First, it appears as if global pollution may well become serious enough in three or four decades to cause sufficient global chilling to diminish crop production significantly. Secondly, long-term projections of the availability of fossil fuels typically overestimate the time to depletion, because of insufficient attention to the rate of buildup in demand per capita for oil and gas in the developing countries. Oil and gas will be just about depleted by the turn of the century, globally, in a trouble-free scenario.

A third forecast will be of particular interest, because it is going to be tested in the very near future. Demographers have trouble in agreeing about whether the recent sharp drop in birth rates in the United States is a temporary phenomenon, to be reversed in the next year or two, or is more permanent. Projections from environmental systems modelling indicate, as mentioned earlier, that the drop in birth rates is largely explained by unemployment in the young, brought about by an 'excess' of fourteen- to twenty-four-year-olds. Further, since the number in this age-range is going to increase, the birth rates will continue to drop for at least ten years. This trend will clearly have great impact on the plans of those who keep banking on a rapidly expanding economy and develop housing, for example, on that basis.

Certain broad, general conclusions have been arrived at by several teams of environmental systems modellers. Holling and his co-workers at the University of British Columbia, Donella and Dennis Meadows at the Massachusetts Institute of Technology and my group at Davis have all demonstrated that, as the human population approaches various types of limits to growth, by saturation or depletion, changes will come at an increasingly rapid rate. Since many different types of changes, most of them traumatic, will be occurring simultaneously, they will severely tax the ability of archaic and slow-moving institutions to respond appropriately. We obviously need new institutions, with faster response characteristics.

We are also getting close to the end of the period when 'economic growth' will appear to be a viable social driving policy, even to the most gullible and naive. Some big surprises are in store over the very short-term, particularly with regard to shortages of critical fuel and minerals. All of these systems-modelling groups show, as do others, that solutions to our problems will be effective only if much more efficient use of resources is coupled with population control. Either without the other will buy us little time.

The current situation

Two possible strategies for systems modelling have now been tried. One is to focus effort on building a tremendous amount of detail, and hence realism, into all systems subcomponents, then to hook these together and attempt simulation of the entire system. This approach runs into two problems: no one can understand the output, because of its volume, and each computer run of the entire system costs too much.

The second approach is to try to build models that deliberately sacrifice detail, but concentrate on incorporating the essential driving mechanisms and key variables into the model. The kind of understanding required for judicious selection of driving mechanisms and variables may only come after detailed work with models of individual system components, or small groups of system components. This second approach leads to output comprehensible to the human mind, and in which the essential aspects of the system are revealed. Also, individual runs of the whole system model are not too expensive, and the consequences of many different policies and combinations of policies can be explored with a reasonable budget.

In short, while we need detailed models to answer specific questions about system components, we need highly aggregated models to understand the dynamic behaviour of the whole system.

References and further reading

J.M.Dutton and William H.Starbuck (eds.), *Computer Simulation of Human Behavior* (New York and London 1971). See particularly chapters 27 and 28, which describe models of the United States economy.

Jay W. Forrester, *Urban Dynamics*, Massachusetts Institute of Technology (Cambridge, Mass. and London 1969).

Jay W. Forrester, *World Dynamics* (Cambridge, Mass. 1971).

D.H.Meadows, D.L.Meadows, J.Randers and W.W.Behrens III, *The Limits to Growth* (New York and London 1972).

Bernard C.Patten (ed.), *Systems Analysis and Simultion in Ecology*, 1 and 2 (New York and London 1971 and 1973).

Kenneth E.F.Watt, *Ecology and Resource Management* (New York and London 1968).

Kenneth E.F.Watt, 'A model of society. Simulation', *Technical Journal of Simulation Councils*, 14 (1970), pp. 153–64.

Kenneth E.F.Watt, *Principles of Environmental Science* (New York and London 1973).

AFTERMATH OF THE GREEN REVOLUTION

Gordon Conway

A hundred years ago the upland people of Thailand were able to support themselves through the practice of shifting cultivation. They felled and cleared the forest, grew a sequence of crops and then moved on, leaving the land fallow to restore its fertility. As a livelihood it provided no more than subsistence; but it placed little demand on the local environment and hence was relatively assured from year to year.

Tropical soils are often infertile. Centuries of heavy rains have leached out the salts and minerals and the profusion of plant and animal life, which so impresses the visitor from temperate countries, depends on a fragile cycle of nutrients. The shifting cultivator is aware of this. He knows that the length of the cropping period and the fallow after it are critical. The land needs to be left for at least ten years after cropping before the fertility is fully restored. If it is cropped again too soon, yields are poor and the soil is quickly exhausted.

Today in Thailand high birth rates coupled with improved medical care are increasing upland populations by 3.5 per cent or more each year and the people are seeking higher standards of

Gordon Conway is a research fellow in the Environmental Resource Management Research Unit of Imperial College, University of London, where he works on mathematical models of pest control. Since working in Borneo 1961–6, he has continued to study South-East Asian problems for the Ford Foundation.

living. Under these pressures the traditional cycle of shifting culti-vation is breaking down. Yields are deteriorating, soils eroding and large areas of land becoming abandoned. In Northern Thailand three to four million hectares of watershed are affected in this way.

The effects are not simply confined to the uplands. As the soils erode they lose their ability to absorb and store water. Thus when the rains come they wash straight off the hills and the rice crop, which needs a steady, sustained supply of water, suffers. Flooding too becomes more serious and the heavy burden of silt that is carried down fills lakes and reservoirs, chokes irrigation canals and alters estuaries and swamps, affecting navigation and fisheries.

The plight of the shifting cultivator demonstrates the more general problem. How can human needs – adequate food and clothing, a decent place to live and freedom from serious illness – be satisfied without the attempt destroying the basis upon which such development eventually depends? All environments exhibit a conflict between productivity and stability. The tropical forest is highly diverse and hence stable; yet it is not very productive. Development entails a shift in the natural balance, a process of creating simpler, semi-artificial environments in which the produc-tivity is higher and more useful. Cropped fields replace the forest; cities grow where villages stood. The perennial hazard is that serious instability may arise from this process of environmental simplification.

The cotton farmer in the coastal valleys of Peru operates on a more developed and, by Western standards, more sophisticated pattern of agriculture than the shifting cultivator of Thailand. In the Cañete Valley, for example, the farmers have for many years run their own experimental station and after the Second World War they were among the first to use new organochlorine insecticides, such as DDT and BHC, to simplify their control of crop pests. At first cotton yields increased; from 350 to 450 kilograms per hectare in the later 1940s to nearly 600 kilograms per hectare by 1954. But the valley is surrounded by dry country with little vegetation, which effectively isolates the pest populations; the insecticides were being sprayed from the air to give a blanket cover of the valley and this combination of conditions produced an intensive selection pressure on the pests so that resistance developed rapidly. In addition populations of the more susceptible beneficial insects, parasites and predators, were wiped out. The original pests increased and new ones appeared. By 1956 yields had dropped below 300 kilograms per hectare.

The response of the cotton farmers to this disaster was a radical change in their approach to pest control. Instead of relying on blanket application of organic insecticides an integrated control programme was developed which emphasized careful timing of cultivation practices, importation of beneficial insects from neighbouring valleys and use of only those insecticides which are selective in action. This policy, which amounted to restoring some complexity to their system, was highly successful and by 1960 yields were averaging 750 kilograms per hectare.

The Green Revolution

For the recent introduction of new varieties of cereal crops throughout much of the tropical world the name Green Revolution is no exaggeration. In its initial phase, it represents the most far-reaching simplification of the global environment ever attempted by man.

In the 1940s the Rockefeller Foundation began a co-operative programme with the Government of Mexico aimed at improving Mexican food production, principally by developing new varieties of wheat and corn. At the time average yields of wheat were but a quarter of those obtained in the United States. The plant breeders first worked on developing resistance to rust disease; but then in the 1950s they decided to aim for a quantum jump in yields by attempting to perfect a plant with a new kind of structure and physiology. They had found that the local varieties were liable to lodge (fall over) if subjected to heavy applications of fertilizer. A plant with a shorter, stiff straw could absorb the fertilizer without lodging. They also wanted a wheat that would mature more rapidly. By 1961, through planned crosses and careful selections, they had several new varieties which met these specifications. Yields of over 7,000 kilograms per hectare of grain were now possible. The new varieties were rapidly adopted and average wheat yields in Mexico doubled in the next ten years. Many other countries are now benefiting from the improved wheats.

This achievement with wheat and similar success with corn encouraged the setting up of a parallel scheme for rice improvement in Asia. In 1962 the Ford and Rockefeller Foundations and the Philippine Government jointly established the International Rice

Research Institute at Los Baños in the Philippines. With the Mexican experience behind them the plant breeders had a clear idea of their objectives and the approach to be followed. Success was rapid. By crossing a tall, vigorous variety from Indonesia with a dwarf Taiwanese variety and selecting from the best of the hybrids, they produced a new rice which they named IR-8. It was short and stiff-strawed; it matured in 120 days as opposed to the 150–80 days of the local varieties; and it could absorb up to 110 kilograms per hectare of fertilizer giving yields of over 9,000 kilograms of grain per hectare. IR-8 was released in 1966. By 1968, thirty million hectares in Asia were planted to this and other high-yielding varieties, and in the same year the Philippines became self-sufficient in rice production for the first time.

The new wheats and rices have been called 'miracle' varieties. But it is not so much what the plants can do that is remarkable; colonial agriculture had shown that greatly increased yields could be obtained in the tropics from crops such as rubber, oil palm and sugar cane. The miracle lies in the speed with which the new varieties have been produced and put into the farmer's hands. And because of this speed and the revolutionary nature of the development, problems have begun to manifest themselves.

The varieties are all derived from narrow genetic stock. They replace numerous local varieties, selected over centuries for their suitability to local environments and in particular for their resistance to a wide range of potential pests and diseases. The danger is that new pests and diseases will arise and, because of the continuous nature of the cropping, will explode into major outbreaks. These are not groundless fears: in many countries the new varieties have already shown themselves vulnerable to attack. One disease that has become particularly important is bacterial leaf blight of rice. However the most severe problem has been caused by increasing populations of rice leafhopper, an insect which carries tungro virus. In Bangladesh this virus has completely prevented the growing of IR-8 and in the Philippines a major outbreak in 1971 was one of the principal causes for a large drop in rice production which has brought the country once again below the level of self-sufficiency.

Pests and diseases are only one set of 'second-generation' problems of the Green Revolution. To attain high yields with the new varieties requires not only seed but other inputs as well – heavy applications of fertilizer, an assured and controllable water supply and effective weed control. To buy seed, fertilizers and herbicides requires capital; the costs may be a hundred times those involved in growing the traditional varieties. In the Philippines the small farmer seems, by and large, to have been successful in obtaining the necessary credit; landowners have become wealthy and there has been real growth in the income of tenant farmers. Elsewhere, in parts of India and Ceylon for example, the benefits have been less evenly divided. The bigger landowners have had preferential access to the inputs required and have prospered at the expense of the smaller farmer. With more capital available they have turned more readily to mechanization and this has increased rural unemployment. In Tanjore, one of the first agricultural districts in India to receive the new varieties, serious rioting broke out in 1968 over the sharing of the new wealth.

There have been difficulties too with marketing and distributing the grain. But on balance the benefits have outweighed the problems. There is no justification for arguing that the new varieties should never have been developed. The questions now are about future directions: how should the problems be tackled and how can the obvious potential of the new varieties be harnessed to create a permanently productive agriculture in the developing countries?

Exchanging yield for stability

In recent years integrated pest control programmes, like that adopted by the cotton farmers of the Cañete Valley, have been successfully developed for a number of tropical crops – cocoa and oil palm in Malaysia for example. An attempt is now being made to produce such a programme for rice. In Southern India there exists a local variety of rice with a wide spectrum of resistance to pest and diseases. At the International Rice Research Institute this has been crossed with high-yielding varieties to produce a new rice, IR-20, which has inherited much of the resistance. The next stage is to devise techniques of monitoring pest populations which will ensure that insecticides are only applied when and where they are really needed. In this way the development of insecticide resistance can be slowed down and there is less risk of affecting natural parasites and predators.

The development of IR-20 has a significance beyond its role in obtaining better pest control. Underlying the programme which resulted in the early varieties, such as IR-8, was a philosophy of plant breeding that placed an almost exclusive emphasis on producing very high yields and income under ideal conditions. But the goal of the average farmer is more complicated than this. Because of the cost of inputs a strategy which concentrates solely on maximizing return in this way inevitably involves a high risk. If the weather is bad, or the irrigation fails, or pests and disease strike then the result may be a net loss rather than gain. Tropical climates vary greatly from one year to the next. 'Average' conditions rarely happen; in the Indus Valley, a review of the climate for the past hundred years has shown that bad weather occurred one year in every three, droughts and floods with equal frequency.

The small farmer knows this and takes it into his calculations. His goal is not to maximize net income in any one year but to obtain, for a certain range of expenditure, a good income taking one year with the next. The kind of crop variety he needs is one which will yield well in average years, respond with higher yields if the weather is good and the inputs available, but in bad years will still produce a reasonable return. The new variety, IR-20, is a step in this direction. Not only has it broad resistance to pests and diseases but it is tolerant of poor soils and less susceptible to mineral deficiencies. It does not perform as well as IR-8 under ideal conditions; it lodges when given very heavy levels of nitrogen, but it is less likely to do worse than the traditional varieties when conditions are not so favourable.

Developing broadly adapted crop varieties is part of the answer, but the more fundamental ecological solution to the problem of instability and risk lies in diversifying cropping systems. A common feature of traditional farming in the developing countries is the extent to which it relies on a mosaic pattern of cropping, different crops being grown on the same land, either together or in some form of rotation. Such systems have many basic advantages. They make much greater use of the available sunlight. By including legume crops, such as soya beans, they can help to maintain soil fertility and cut down on the requirements for chemical fertilizers; and by including crops such as sorghum or mung beans, they can provide an insurance against drought. There is also a potential for pest and disease control, the more diverse environment favouring the activities of beneficial insects.

In parts of Java the growing of five or six crops a year has been practised for centuries. Such cropping systems have been regarded as museum pieces, but this attitude is now changing as their strengths and potentialities are being realized. At a number of centres in the tropics, agronomists are beginning to develop modern versions of multiple cropping which depend on better techniques of cultivation and the judicious use of the new grain varieties (see table). The agronomic work is being combined with detailed economic studies so that cropping patterns can be devised which will

Yields from a highly intensive experimental multiple cropping system at the International Rice Research Institute. The land is cultivated into ridges and furrows, rice is grown in the furrow, sweet potatoes on the ridge and so on through the cycle (after Bradfield 1972)

Crop	Planted	Harvested	Days	Yield Crop	(tons/ hectare) By-product
Rice	1 June	30 September	120	5.0	5.0
Sweet potato	1 September	24 December	114	25.0	20.0
Soya beans	27 December	17 March	85	2.5	—
Sweet corn	1 March	5 May	66	40,000 (ears)	15.0
Soya beans (green pods)	1 May	1 July	60	6.0	6.0
Rice	1 June	etc.			

satisfy local marketing demands, pressures for greater employment and the elements of risk particular to different regions. If the potential inherent in this approach could be fully realized the benefits would be enormous. Multiple cropping could help stabilize and improve rural life, free capital resources for industrialization and at the same time stem the migration from rural areas.

Regional land-use and home-grown technology

With multiple cropping we have a viable approach to productivity on the farm. To make policy and create sound environmental plans on a regional scale poses a bigger challenge. At present land-use policy in developing countries tends to be formulated on the basis of simple checklists of economic and political priorities rather than with the aim of achieving some optimal productivity for the whole land system. Frequently policies emerge as the result of unrelated and often contradictory decisions.

But even when planning is highly sophisticated, as for example in Malaysia, the environmental perspective may still be weak. Malaysia's land-capability classification, which is the basis of their extensive rural development programme, lists potential mining land as class 1, agricultural lands as classes 2 and 3, productive forest as class 4 and protection forests and unproductive land as class 5. The hierarchy, here, determines the eventual use. Thus land with mining and agricultural potentials is allocated to mining, productive forest with agricultural potential to agriculture and so on. Protection land is merely a residual. Yet the conservation of upland soils and vegetation is essential for the continued productivity of lowland areas. If the residual that is left for protection is too small, after all other uses take their share, then the total productivity of the landscape is likely to decline.

The fault does not really lie with the planner; he can only act on the advice and information he receives. Perhaps the major challenge of environmental science is to provide information in such a way that it is commensurate with that from other sources and is comprehensible to planner and higher policy-maker alike. The setting aside of conservation land, for example, is only likely to receive serious attention when it is possible to define fairly precisely for each region the siting and size of the area required, the most suitable pattern of agriculture or forestry and the kind of management that is needed.

The problem arises acutely in the planning of dam and reservoir projects. At present feasibility studies concentrate on the hydroelectric potential, the costs and logistics of the dam construction, and so on. The techniques for obtaining and analysing this kind of information are well known; by contrast it is far less easy to forecast the effects of the dam on the fisheries or on the spread of water-dwelling snails that can carry disease (bilharzia). In part, it is a lack of basic ecological data: what determines fish production in lakes; how rapidly do disease organisms spread? Nevertheless some of this information is available; it needs to be put together in a more meaningful manner. Environmental science has to adopt some of the techniques, such as systems analysis and computer simulation, which engineers have so successfully used. We are at a stage where mathematical models could be built of reservoir fisheries or of the dynamics of bilharzia and other diseases. The results would be crude, admittedly, but if they were tied in with management and refined by experience they could begin to provide information of the kind that planners require.

Cities are ecosystems in the same way as cropped fields, forests and river basins. Although the elements are more physical than biological and the systems are more dominated by human activity, we can find the same perennial conflict between productivity and stability. Poor management of an upland area results in pests and diseases, soil erosion and poor crop yields; in the city the results are congestion, pollution, disease, crime and related social problems. The remedies lie in methods of urban management that are in some way analogous to integrated pest control and multiple cropping.

A new approach to urban sewage treatment in the developing countries gives a sense of the kind of techniques that are needed. Engineers at the Asian Institute of Technology in Bangkok have shown that it is possible to utilize the strong sunlight of the tropics to promote rapid bacterial breakdown of sewage and at the same time produce rich algal harvests. The algae can be piped into fish ponds or dried and fed to pigs or chickens while the remaining water can be readily purified for human consumption.

The growth of modern western-based technology has made use of an increasing degree of specialization – a form of educational simplification. As a consequence few people today have a training broad enough to understand the fundamentals of the variety of disciplines that environmental science requires. The kind of expertise that is impressive and carries weight in the West tends to be that which comes from long experience in a narrow field. All too frequently experts brought into developing countries to advise have little ability to fit their own knowledge and judgement to the local environment; and too often local scientists and planners have received an academic or specialized training abroad which ill-equips them either to adapt techniques from the industrialized countries or develop new ones appropriate to their own environments.

Moreover, appropriate technologies do not arise as a simple compromise between the most primitive and the most sophisticated. Techniques such as integrated pest control or recycling sewage combine very advanced with highly traditional elements, often in a skilful and complex manner. They require an awareness of the limits and potentialities of local environments, which comes from intimate experience. A major need, therefore, is far more aid for training and research in environmental science in the developing countries, at universities, colleges of technology, agriculture and forestry schools which are part of the local environment and can benefit from day to day experience.

On the other hand the development of such technologies also requires a high level of analytical skill and innovative capacity. The dominant philosophies in the developing countries, particularly in Asia, differ from those of the West in their view of the relationship between man and nature. They see man more as a participant in natural environments, rather than their overseer. The West, attempting to cope with its current environmental problems, has much to learn from this viewpoint but, in Asia itself, it is not a philosophy that provides a good defence against the incursions of Western technology or the pressures of population growth. The incentives for analysis and control of the environment are weak. Nevertheless the ideals of harmony and integrity with nature provide a basis for a will to better environmental management. Analytical science welded to such a will could provide a powerful force for achieving lasting prosperity in the developing world.

References and further reading

R. Bradfield, 'Maximising food production through multiple cropping systems centred on rice', *Rice, Science and Man*, International Rice Research Institute (Los Baños 1972).

L. R. Brown, *Seeds of Change* (London and New York 1970).

G. R. Conway, 'Better methods of pest control', *Environment; Resources, Pollution and Society*, ed. W. W. Murdoch (Stamford, Connecticut 1971), pp. 302–25.

G. R. Conway and J. Romm, *Ecology and Resource Development in Southeast Asia*, Ford Foundation (New York 1972).

M. T. Farvar and J. Milton, *The Careless Technology* (New York 1973).

M. G. McGarry and C. Tongkasame, *Water Reclamation and Algae Harvesting*, J. Water Pollution Control Federation (May 1971), pp. 824–35.

J. Romm, *A Survey of Urbanisation in Bangkok*, Ford Foundation (New York 1973).

THE ESCAPE
FROM SMOKE
S. Fred Singer

A litre of gasoline produces energy equivalent to more than six man-days of physical work, making the cost of a man-day only pennies. This simple calculation brings home the importance of energy to everything we do. Another illustration of the subsidy we receive from the fossil fuels – coal, petroleum and natural gas – is that the use of energy in the United States today is just about a hundred times the minimum energy required, in the form of food, to sustain human life.

The correlation between per-capita use of energy and each nation's level of economic development is very striking; there is a direct relationship between energy consumption and average income, with the cost of energy itself being only a few per cent of the income. Of the two million megawatts of energy consumed in the United States, about one-quarter goes into the generation of electricity and three-quarters is used directly for transport, heating and other purposes.

In terms even of the brief existence of man on this planet, the energy revolution has occupied an incredibly short span and it would all be over in a couple of centuries if we had only the fossil

S. Fred Singer is professor of environmental sciences and a member of the Center for Advanced Studies at the University of Virginia, Charlottesville. He is a physicist who has previously served as dean of the School of Environmental Sciences at the University of Miami, as director of the National Weather Satellite Center and as Deputy Assistant Secretary of the United States Department of the Interior.

fuels. Even if we add forms of fossil fuel that do not at present count as resources, such as tar sands and oil shales; even if we use underground nuclear explosions to extract gas and oil more efficiently and completely, we are still left with a finite amount of fossil fuel. Much has been made of this fact, including recent demands that we drastically slow down our use of fuels and effectively stop economic growth. Yet, even if we limited consumption, or were in fact to decrease our current use of fossil fuels, we should still run out of them after a certain number of years.

Fortunately, essentially inexhaustible energy sources are in sight: nuclear breeder reactors and (less certainly) nuclear fusion reactors, as well as the possibilities of exploiting solar power more directly and continuously, and of tapping the internal heat of the Earth. The scientific discoveries and technological developments on which the nuclear 'rescue' is based have depended on a high level of economic prosperity in some countries, which allowed us to make the investments in capital and human effort that were required. It is interesting to speculate on whether parsimony in the use of fossil fuels from the outset, to make them last perhaps a thousand years, would ever have created the conditions in which, in our profligate way, we have won our long-term independence.

The energy revolution has produced not only spectacular advances, but also spectacular problems. It has created for about one-fifth of the world's population an undreamed-of standard of living. But advances in medicine and agriculture have produced unprecedented population growth as well – first in the developed countries, but now also in the underdeveloped nations of the world. The world population stands at 3.5 billion and will double again in the next thirty-seven years or so. So far, food supply has kept pace, but just barely, and then only because of the subsidy from fossil fuels. Tractors and other machinery are replacing draft animals and human labour; factory-made fertilizer augments the natural nutrients of the soil. In that sense, bread and meat are made partly from oil!

With population growth has come a greatly increased consumption of resources – and also of pollution, both locally and worldwide. Man's activities are now making a measurable impact on the global environment – an impact which will increase as world population grows and reaches even higher levels of consumption. It is my purpose to discuss this other face of energy.

Products of combustion

If fossil fuels were chemically pure and were completely burnt the only direct products of their combustion would be water vapour and carbon dioxide. No one has yet suggested that water vapour is a significant environmental hazard, except conceivably at the high altitudes where supersonic aircraft operate. Carbon dioxide, like water vapour, is a natural constituent of the atmosphere and it is not poisonous in the ordinary sense. The possible environmental effects of carbon dioxide released from burning fuels are of different kinds, and are considered by other authors (see especially MacIntyre, Chapter 14, *editor*). In the context of this chapter, I shall remark only that carbon dioxide and its possible effects on climate is a transient problem, because of the impending exhaustion of fossil fuels.

The common pollutants that combustion injects into the air include sulphur oxides, a product of chemical impurities in the fuel, together with carbon monoxide, hydrocarbons and solid particles, which result from incomplete burning, and also nitrogen oxides, formed because nitrogen of the air becomes extraneously involved in the combustion process. Many other chemical compounds occur in trace quantities in the output from furnaces and engines. In addition to the products of combustion which are quasi-natural, in the sense that a forest fire produces them too, lead is present as an impurity deliberately added to gasoline to improve its performance.

The principal sources of air pollution in the United States 1968 (Extracted from SCEP 1970)

Pollutant	Mass released (million short tons)	Principal sources (contribution per cent)
particles	28	non-fuel sources (64)
sulphur oxides	33	coal furnaces (60)
hydrocarbons	32	gasoline vehicles (47)
nitrogen oxides	21	gasoline vehicles (32)
carbon monoxide	100	gasoline vehicles (59)

The table above shows the principal sources of air pollution in the United States. It is noteworthy that, in spite of our mental associations with smoke, the burning of fuel is a relatively small source of particles, compared with the combined output from accidental fires, from paper and flour mills, from quarries and similar industrial sources. Even coal-burning, which accounts for nearly 90 per cent of the smoke released from all fossil fuels in the United States, puts no greater mass of particles into the air than cement works do. Of course, more people live near coal furnaces than near cement works, and 'smokeless' forms of coal should be used where possible to reduce the emission but, in relation to global pollution by particles, coal fuel is not the prime offender.

On the other hand, coal is by far the greatest source of sulphur oxides. These gases are harmful in a variety of ways, directly attacking the respiratory systems of humans and animals, injuring plants, and forming acids that damage buildings, textiles and machinery. Depending on weather conditions, the effects of sulphur pollution can be perceptible hundreds of miles from its origin. Globally, sulphur emission gives cause for concern partly because the man-made emissions seem to be of the same order of magnitude as emissions from volcanoes and other natural sources, and partly because the eventual fate of the sulphur is uncertain. Oil-burning furnaces, too, are a significant source of man-made sulphur pollution. Control of sulphur, by treatment of the fuels or of the vented gases, is plainly one of the chief measures of air-pollution control that industry will find itself under increasing pressure to adopt. This is important as coal reserves will last considerably longer than oil, and sulphur is the chief pollutant from coal.

For all of the other major air pollutants, one machine stands out as the prime offender: the gasoline-driven motor vehicle. The high concentrations of carbon monoxide attained in New York streets in the rush hour, and the photochemical smog of Los Angeles and other cities produced by the reactions of hydrocarbons and nitrogen oxides in sunlight, are too notorious to be worth describing. When one adds the pollution of both air and water caused by the extraction, transport and processing of petroleum, largely to meet the demand for gasoline, this turns out to be an exceptionally polluting form of transport. The direct effects of automobile pollution, though, tend to be concentrated in urban areas, and globally the man-made hydrocarbon vapours, nitrogen oxides and carbon monoxide are probably insignificant. The spillage and dumping of oil, although extensive and at times conspicuously fatal to marine life, seems so far to be non-cumulative in its effects, suggesting that the oil is eventually degraded by bacteria or other organisms. Without condoning the careless handling of oil, one should not forget that the environment has often had to cope with natural seepages of petroleum and gas.

To return to air pollution directly caused by automobiles: regulations coming into force in the United States and other countries are intended to check the increase in noxious exhausts and eventually to reduce them. Redesigning the well-proved internal-combustion engine to meet higher standards of pollution control will not be easy; for one thing, there is a conflict between the need to reduce nitrogen oxides, which means reducing the maximum temperature attained in the engine, and the wish to achieve more thorough combustion in order to burn up the hydrocarbons and carbon monoxide. There is interest in electric and other unconventional motors for private vehicles, but so great has been the technological investment in the internal-combustion engine that the innovations are unlikely to compete successfully by the ordinary economic tests. On the other hand, air pollution is only one of the factors that is driving urban planners to look for a revival in public transport and the days of the conventional automobile are in any case numbered, because of the impending exhaustion of the world's oil.

Electricity is rightly seen as an exceptionally clean form of energy, at its point of use, but of course it has to be generated somewhere, usually by burning fossil fuel or, increasingly, nuclear fuel. Only hydro-electric power is completely pollution free, but it entails other environmental problems including competition for scarce water resources and the ecological effects of dams, reservoirs and transmission lines. Moreover, as people live on plains and water cascades on mountains, there is a basic mismatch, both regional and global, between the sources of hydro power and its potential users. Advances in long-distance electrical transmission may eventually allow us to tap the hydro power of the Himalayas and the Andes, but that would depend as much on settled international relations as on technology.

Geothermal and solar power

Geothermal power, obtained from the internal heat of the Earth, is in somewhat the same position as hydro-electric power. At present it can be obtained economically only in a few places around the world, chiefly in New Zealand where it produces a large fraction of that country's electric energy. There is hope that we may learn how to tap heat energy even from volcanoes at an economical cost and with safety, and thereby increase substantially the world supply of energy. The technology of tapping geothermal energy resembles superficially the technology for extracting oil. Holes are drilled into the ground and steam is obtained to run electric generators. In cases where natural steam is not available it is necessary to pump down water for turning into steam which is then used above ground.

Another potential geophysical (or rather astrophysical) source is solar power. Of course, nearly all of our energy derives from the Sun, including the food we eat and the fossil fuels which represent solar energy stored up over the past hundreds of millions of years. But up till now the direct collection of solar radiant energy and its conversion into electricity has not been economical. Nevertheless, there is every reason to expect that large-scale installations may become economically competitive as their design improves and as the cost of conventional energy sources rises. Here again, as for hydro-electric and for geothermal power, there may be certain regions of the Earth, such as desert areas at low latitudes, which would be most suitable for the collecting and converting of solar power. It seems likely that it would not be used directly as electricity but stored in intermediate forms, for example as ammonia or liquid hydrogen, to be shipped to places around the world where energy needs are greatest. Simple calculation shows that the energy needs of the Earth could be met by collecting the solar energy falling on 0.01 per cent of the Earth's surface. Still, this is a rather large area.

Nuclear energy

Discoveries and inventories from all the sciences are cumulatively of great value in allowing us to see precisely what our potential sources of energy are. It is this global view that shows the long-term importance of nuclear power technology, as the only substantial compact source of energy left to us, beyond the fossil fuels. In the twenty-first century, nuclear reactors generating electricity will be the prime source of energy for mankind, although the practical advantages of liquid fuels, so clearly demonstrated in the age of oil, will no doubt be retained by using nuclear energy to make energy-rich, pollution-free liquids.

For existing forms of nuclear reactors, fuel supplies are strictly limited. They depend on the fission of uranium-235, a relatively rare component of uranium; even though a gram of U-235 is worth nearly three tons of coal, the cheaper supplies of uranium are likely to be exhausted by the end of the century. The first major improvement in the fuel-supply situation is expected from the 'breeder' reactors now at an advanced stage of development. These will convert much commoner forms of the heavy elements (uranium-238 and thorium-232) into nuclear fuels, at a faster rate than they burn the fuels. This technology will open a truly vast storehouse of energy in the Earth's crust, far exceeding the fossil fuels.

The catch is well-known, that all fission reactors, whether conventional or the new 'breeders', produce a mass of radioactive material as waste products. This means that great precautions must be taken, in design and operation, to avoid accidents that might rupture the pressure vessels of the reactors and release large quantities of radioactivity into the environment. A few minor accidents have occurred and, as the number of nuclear power plants grows, there is understandable anxiety that somewhere, sometime, there will be a major disaster with a nuclear reactor, perhaps forcing the evacuation of a densely populated area. Wars or sabotage could greatly increase the dangers. From an environmental point of view, however, the choice is between continuous, world-wide pollution from coal-burning power stations, and a finite but very small risk of acute local pollution in the event of a nuclear accident.

Radioactive wastes have to be routinely removed from the reactors, transported, processed and stored safely. When mankind is meeting a large part of its growing energy needs from nuclear fission, the amounts of radioactivity will be enormous. There is little difficulty, in principle, either in keeping the wastes in failsafe tanks indefinitely or in burying them at great depths in carefully chosen rock strata where the chance of their ever escaping is effectively zero. The costs of such storage will grow, however, and be reflected in the unit-costs of electrical power. According to one estimate, by the year 2000 the United States will be storing about 180 million litres of high-level liquid waste, containing some 5,000 tons of fission products.

In addition, all nuclear operations, and predominantly fuel reprocessing, produce water or air contaminated with diluted radioactivity, so that those could never be economically retained; this low-level waste eventually finds its way into the oceans. Of special concern are radioactive elements among the fission products which decay relatively slowly and are chemically concentrated in living tissue and the food chains of animals, including man: caesium and strontium are the most notorious of these. Given internationally agreed standards of disposal, such as those suggested by the International Atomic Energy Agency, there is no reason why it should be allowed to become a problem, although some nations may try to cut power-generation costs by dumping excessive amounts of radioactivity.

A current exception to this sanguine generalization is krypton-85, a radioactive gas which is peculiarly difficult to trap and control during fuel reprocessing because it is chemically inert. At present it escapes into the air. Methods of containing krypton-85 are under development, but how successfully remains to be seen. In addition, plutonium itself, the key to the breeder-reactor fuel cycle, is particularly hazardous and special care will be needed to prevent its escape into the environment.

Radioactivity is the easiest type of pollution to monitor, even at very low levels, because individual atoms are readily detectable. The same is true for natural radioactivity and cosmic radiation, present everywhere on land, which provide an unusually clear standard of comparison for man-made radiation hazards. Because radiation can have more fundamental biological effects than most known pollutants – damaging, among other things, the genetic material – any addition is unwelcome. It seems entirely possible that the radiation burden to the human population due to nuclear power generation can be kept down to a thousandth part of the average natural radiation, which is far less than the variation in natural radiation from one environment to another.

Beyond the breeder reactors lie the thermonuclear fusion reactors. While fission reactors depend upon the break-up of some of the heaviest elements, fusion reactors will derive energy from the combination of light elements into somewhat heavier elements. In the process, some radioactive materials will be formed, but on a small scale and as incidental by-products rather than as large-scale products of the basic nuclear process, which is the case with fission. This supposes that practicable fusion reactors can be built, and for two decades there have been alternating hopes and pessimism about the prospects of achieving the extremely high temperatures needed to maintain the reaction. Fortunately, the breeder reactors will give us breathing-space to find the technical solutions, although plainly the sooner fusion can supersede fission the less anxiety will there be about the accumulation of man-made radioactivity.

The lesser temperature target is that for the tritium-deuterium reaction, in which two heavy forms of the lightest element (hydrogen) fuse together. Deuterium (hydrogen-2) is very common, comprising one part in 4,500 by weight of sea water, whence it can be extracted without great difficulty. Tritium (hydrogen-3) is virtually non-existent in nature and will have to be made as it is needed by the fission of lithium-6. This material is, if anything, less abundant than uranium, but there is enough of it in the Earth's crust to contribute to power generation for many centuries. The true liberation of mankind from any misgivings about future power supplies (and the pollution associated with them) will come when technologists achieve the higher temperature target for practical fusion reactors, at which deuterium will react with itself. When that stage is reached, the energy potential of every litre of sea water will be equivalent to several litres of gasoline.

The only remaining pollution problem will then be waste heat, which can raise the Earth's temperature locally or globally. If one assumes – optimistically or pessimistically, according to one's point of view – that in fifty years the rest of the world will reach the present United States level of energy consumption, and that the population will then be three times the present population, the total power consumed would be 110 million megawatts. The Earth radiates solar energy back into space at a rate of more than a thousand times greater than that. It would be rash to assume, however, that man-made heat put into the environment can therefore be neglected. The energy would be distributed in a patchy manner, reflecting the local concentrations of population and the circulation of air and water. In particular places waste heat may have far-reaching consequences either on the climate or on living organisms directly exposed to the heat. Thermal pollution is thus one aspect of power production that will require careful attention, so long as man concentrates energy to do his work for him.

Grounds for hope

The recent public 'debate' about the environment abounded in irony for those of us who had spent a substantial part of our working lives worrying about the ecological relationships of industrial man, since long before the question became fashionable. From being scientific radicals, eager to reform the wicked ways of industry, we found ourselves driven into a more conservative position by extravagant claims that the world was coming to a horrible end. Nobody knows enough about the environment to be dogmatic about anything, but the more one investigates any issue – energy and air pollution, for example – the less relevant do many suggestions about perils and remedies appear to be to the urgent task of bringing an enlarging human population into harmony with its environment.

The aversion from nuclear energy, conspicuous among many environmental pamphleteers, is a case in point. It is so irrational that one must look for hidden motives. The common origin of power reactors and nuclear weapons is no doubt one explanation, combined with fears engendered during the period of unrestricted nuclear tests. Less forgivable is the determination to see no good of any kind coming from technology. An unprejudiced assessment of nuclear energy shows it to be man's best hope for reducing pollution, provided only that the control of radioactivity is as stringent as everyone concerned knows it must be.

An increase in world prosperity, especially to meet the needs of the world's poor, argues for more energy use. Environmental considerations, of resource conservation and pollution, would argue for less. The dilemma is resolved in the short-term by spending a small portion of the energy subsidy on pollution control; in the long-term by developing the technology for inexhaustible and truly clean sources of energy. The picture looks hopeful.

References and further reading

Committee on Resources and Man, *Resources and Man* (San Francisco 1969).

P.R.Ehrlich and A.H.Ehrlich, *Population, Resources, Environment* (San Francisco 1969).

SCEP (Study of Critical Environmental Problems), *Man's Impact on the Global Environment* (Cambridge, Mass. 1970).

Scientific American (an issue devoted to the biosphere) 222 (September 1970), 3.

Scientific American (an issue devoted to energy) 224 (September 1971), 3.

S.F.Singer (ed.), *Global Effects of Environmental Pollution* (Dordrecht 1970).

S.F.Singer (ed.), *Is there an Optimal Level of Population?* (New York 1971).

GLOBAL BUDGETS
IN PERSPECTIVE

Michel Batisse

The concept of natural resources includes everything in the material universe which man can use directly or indirectly for what he considers his benefit. Sunlight and air, trees, coal, waterfalls – all these are examples of natural resources. It is a dynamic concept because what is considered useless at one time (say, uranium ore or steep mountain slopes) may become a major resource through a change in technology (nuclear power) or a change in ways of life (winter sports). It is mainly an economic concept, since a potentiality of the natural world becomes a resource only when its utilization is profitable. But it is a useful concept since it reminds us that man is neither angel nor demigod and entirely depends for his survival on the resources he can draw from his environment. The relationship between man and these resources has never been an easy one and it would be difficult to decide whether the future of mankind is more precarious today than it was during the drastic changes of climate which took place some ten thousand years ago, or at the time of the great plagues and famines which punctuate our history.

What is new – at least in the industrialized world, where religious fatalism has been replaced first by reliance on science and technology and then by increasing mistrust of them – is a greater concern for

Michel Batisse is director of the Natural Resources Research Division of the United Nations Educational, Scientific and Cultural Organization (Unesco) in Paris. In that capacity he has been responsible for the Arid Zone and Humid Tropics Research Programme of Unesco, for the International Hydrological Decade and for the new interdisciplinary programme on Man and the Biosphere.

our future together with the uncomfortable feeling that we are responsible for it. A greater understanding of interactions, and an increased capability to foresee and to plan, contribute to this new attitude and so does the tendency to look at problems on a global scale. At a time when we begin to realize that we have 'Only One Earth', and when communications and transport from one part of the world to the other are so swift and abundant, a global approach to the problem of resources seems at first sight to be the only meaningful one. Furthermore, it can provide a relatively small set of global numbers and estimates which appear to give us a better and simpler grasp of the problem. Nevertheless, serious reservations about the global approach are necessary, as we shall see.

Growing demands

The demand for resources obviously depends on the size of the population, but also on the standard of living of that population and on the available technology, since a more efficient technology can produce more goods, or substitute one product for another. World demographic and economic statistics provide a fairly good set of global figures both on the demand and the supply of resources but current raw figures are of little value in throwing light on the population/resources problem. We are dealing with a dynamic relationship and are more concerned with future prospects than with the present situation, no matter how imperfect it may be. Unfortunately, estimates of future demand and supply are extremely hazardous and it is very hard to make reliable socio-economic forecasts even for a few years in advance. Experience has shown that most quantitative predictions made in the last thirty years have been grossly in error and that futurology (which has been described as the art of continuing curves in the direction in which they seem to go) can be at the same time highly sophisticated and utterly misleading. In a period of unprecedented rate of technological, economic, sociological and cultural change, it is in fact impossible to picture the world one generation ahead.

Yet the anxiety about the future, as well as the need for economic and social planning and forecasting imposed by modern technology itself, compel us to make some estimates, even if we shall have constantly to revise them.

There is no dearth of projections for the population of the world.

The latest United Nations estimate gives 6,500 million people by the year 2000 with a gross imbalance between developed and developing countries. In the recent past, predictions of population have always fallen short of reality. This time fortunately the error may be in the other direction in view of decreasing fertility rates in many countries.

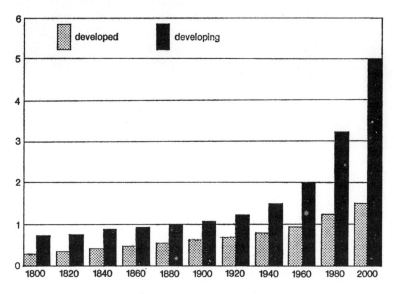

Population estimates for developed and developing countries, in billions (UN data adapted from Ward and Dubos 1972)

As regards the standard of living – a simple expression to describe a wide variety of demands on all types of resources - one can only make two remarks. The first is that the most industrialized countries, which are exerting by far the greatest pressure on natural resources, may reduce somewhat the rate of increase of this pressure between now and 2000; in absolute terms, though, the consumption of resources per capita in these countries will continue to grow. Energy consumption is a rather good indicator of material standards of living. The diagram on the opposite page, derived from UN data and projections, indicates that the developed world has little intention of reducing its use – and waste – of energy.

The second remark is that the developing world will do its best to try to imitate the industrialized countries and to achieve higher and more decent standards of living. The energy diagram assumes meagre success in this direction and the contrast with the population diagram speaks for itself. The continuing disparity between rich and poor nations is probably the most significant and disturbing

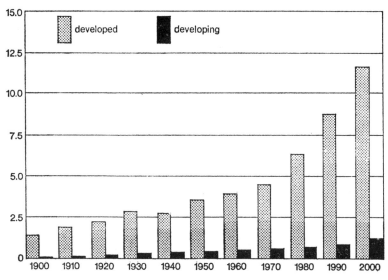

Energy consumption per capita for developed and developing countries, in metric tons of coal equivalent (UN data after Ward and Dubos 1972)

fact in any discussion of population and resources. For the time being, a rough conclusion would be that, in the aggregate, energy demand by 2000 will be at least two and a half times what it is today, and many people feel that it should and perhaps will reach four times the present level.

Another rather crude indicator of the standard of living, the gross national product (GNP) per capita, has rightly been criticized in recent years but it still provides some clues to present and future trends. In the last few years the global growth rate both of GNP per capita and of industrial output has been approximately 7 per cent per year and if this rate were maintained this would lead to both factors being multiplied by four by the year 2000. Here again the global figures obscure the fact that while the rich countries are

249

getting richer, the poor, with their high population increase, are hardly progressing on a per capita basis. For this and for many other reasons – including the environmental awareness which is developing everywhere – it seems doubtful that the present rate of economic growth can be maintained until 2000. The consequence may be that the demand for natural resources for industry will not exceed by then two and a half times the present demand. At the same time, the demand for food and fibre, which is more closely linked to the population figure, should also reach at least two and a half what it is now.

Thus we may adopt, for our conjectures, an overall factor of two and a half for the growth in demand for all resources all over the world in the next quarter-century. It should be stressed that this figure is closer to guesswork than to scientific prediction and that such growth would lead to a world in 2000 not very different from our present one as regards distribution of wealth, with only a start being made towards greater equity among nations. Many different hypothetical worlds can be invented for 2000 and after, including wonderful or horrible ones. This very conservative factor is adopted here because it looks more probable than higher factors while already representing an enormous strain on the planet's resources. The question is, therefore, can supply meet demand, and for how long?

Food – the critical problem

Man's first need is for food, and more than half of the present world population is considered to be ill-fed. In the past decade, despite the first benefits of the Green Revolution, food production per capita in the developing world has shown no overall increase. The land surface of the Earth is huge – more than 13,000 million hectares – but only one hectare in ten is arable land under cultivation. With very considerable labour and capital investment, only one additional hectare out of ten, at present covered by pastures or forests, could conceivably be brought under cultivation. Another $1\frac{1}{2}$ hectares would be left for rough grazing while the rest ($6\frac{1}{2}$ hectares) is not suitable for any significant food production. Opening new land for cultivation will be necessary to cope with population growth in many areas, especially in Latin America and Africa. It cannot go very far in southern and eastern Asia where much less cultivable land is left unused. In any case, the opening of new land is generally a risky and lengthy operation and could not alone meet the food challenge.

More hope is therefore placed on the improvement of productivity on already cultivated lands. In the developed countries, much progress has been made in this respect through fertilizers, pesticides, reduction of storage losses and improved varieties, leading even to cases of local overproduction. Further productivity improvements come up against the law of diminishing returns. In many ways greater progress could be made in the developing countries where the hungry masses are, but there formidable social and cultural difficulties are encountered such as reluctance to accept technical change or transformations in land-tenure systems. Only a slow pace of improvement can therefore be expected. Moreover, the environmental consequences of new cultivation practices, including those of the Green Revolution, have not been fully assessed and call for prudence.

The soil is not, of course, the only source of food and approximately 20 per cent of our protein supply comes from fisheries. The current world catch from the oceans amounts to some sixty million tons and can probably be increased to a hundred million tons by the year 2000. The latter figure, which may be fairly close to the maximum sustainable yield, is well below our overall increase factor of two and a half. It does not justify the high hopes that some people placed in marine resources after the spectacular rise in catch which took place in the 1960s; this rise was largely due to the Peruvian anchovy which, incidentally, is not used directly for human consumption but for feeding poultry and other livestock in developed countries. Other sources of food can be imagined, including proteins derived from petroleum, aquaculture or plankton, but these are not likely to be developed on a large scale as long as conventional sources can still be tapped. The overall picture of the world's conventional food supply has never been bright at any time in history and is not bright today, but it is probably in this area that the limits of the Earth, together with the limits of man's organizing ability, are most constraining.

Water and minerals – room for manoeuvre

The supply of fresh water is related in several ways to the food problem and itself constitutes another vital sector of human needs. Thanks largely to the studies undertaken in the framework of the International Hydrological Decade, the global picture is clearer and countries have both a better understanding of their water resources and a better capacity for handling them wisely. There is a formidable amount of fresh water in the world and figures can be provided for the amount in the atmosphere (relatively small, but with a rapid circulation), in the lakes and rivers (of major practical importance but limited), in the polar icecaps (enormous, but useless so far) and also in the ground, although estimates for the very considerable and slow-moving groundwaters are still quite uncertain.

Comparisons of these various figures are however meaningless in terms of resources since what is important is the actual availability of water where and when it is needed. The vast flow of the Amazon is of no help to irrigate the Sahara. The local shortages of water will remain a limiting factor to development in most arid and semi-arid countries. At the same time, within the general pattern of human settlement imposed by water availability in different parts of the world, the problems of water quality and of pollution constitute a greater difficulty for the immediate future than does the problem of availability. There is, however, little doubt that by stricter pollution control, by rational and multi-purpose utilization of resources, by sufficient capital investment and by increased energy use for transportation or desalination, the demand for fresh water can be met in the foreseeable future.

An area of great concern in many quarters today is that of mineral resources, including of course fossil fuels, partly because these resources are 'non-renewable' and partly because precise and alarming figures about the number of years for which the known reserves of these resources will last at current rates of use are regularly published. For instance the following 'lifetimes' are currently given: 15 years for mercury, 20 years for tin and silver, 25 years for lead and zinc, 35 years for petroleum and copper, 40 years for natural gas. The anxiety is aggravated because these 'lifetimes' are reduced considerably when account is taken of the present exponential growth in demand.

Fortunately, the real situation is not all that gloomy, for a variety of reasons. First the indicated 'lifetime' for known reserves should not be taken too seriously; accurate information on many reserves is not available and these figures are more indicators of possible short-term scarcities than results of exhaustive global studies. Secondly, in the mineral field, spectacular new discoveries are being made which may completely change the outlook. Thirty years ago, for instance, iron-ore supply gave cause for anxiety; there is no longer this concern today. Experience also shows that a slight increase in prices makes exploitation of lower grade ores economical and that a similar result is achieved through improvements in technology: the picture of a fixed stock of a particular mineral being exhausted is graphic but somewhat misleading. Moreover, most minerals could eventually be reused if the scarcity of fresh supplies made their recovery economical, and one mineral may be substituted by another, so that the demand may level off as it already has done for coal. While all wastage of mineral resources should certainly be avoided, and relative shortages will be encountered for specific minerals, it appears that major difficulties in this field can be avoided for the coming decades.

Given sufficient energy, shortages of most other resources can be made good, including water, minerals and even food. Energy resources are therefore considered as the most vital of all. The prospects for energy supplies are discussed by other contributors (see especially Singer, Chapter 23, *editor*). We must keep our fingers crossed about them and especially about the hoped-for benefits of advanced nuclear techniques.

Space, species and amenities

No global projections seem to have been made on the amount of land which is being 'paved' by urbanization and other processes, but this would be a most useful study. The space available on Earth is clearly limited in amount and 'natural' space plays an important role in the functioning of the biosphere, whether by reflecting or absorbing solar radiation, supporting photosynthesis which is the basis of life, or absorbing or evaporating water. These functions are significantly modified by agriculture and even more by buildings, roads and airports. If the global population in 2000 reaches 1.7 times what it is now, the urban population will grow roughly 2.4 times while the rural population will only grow 1.4 times. This trend will be even more pronounced in developing countries, where the forthcoming problems of urbanization appear like a nightmare.

The expansion of man-made landscapes, both urban and rural, is accompanied by the disappearance of wild plant and animal species, of natural areas and other amenities. The great danger is that this process, which is moving very fast, is almost irreversible. There is a clear conflict between the expansion of the man-made part of our environment and the need to conserve the natural part and to ensure the long-term, life-supporting capacity of the planet. The 'capital' of natural beauty and wilderness which we are squandering is hard to quantify but is irreplaceable. The genetic material constituted by wild species, particularly by wild relatives of domesticated species, may be essential for our own survival. No effort should therefore be spared now to protect and conserve what is still left after centuries of neglect and abuse.

This very sketchy analysis of the resources problem obviously fails to cover many aspects which could be of importance for future projections. Some who believe that growth should level out without delay will find it too optimistic, others too pessimistic, and many will rightly feel that it oversimplifies highly complicated issues. It is clearly more a matter of personal judgement than of well-founded objective analysis. But it may give some perspective of our situation in the foreseeable future. It certainly reveals some basic points about global forecasting of demand and supply.

The fallibility of demand estimates

We cannot and probably never will be able to make long-term estimates of future demands of natural resources because we cannot and probably never will be able to tell what the future needs of our descendants will be. Apart from the basic metabolic and shelter requirements of human populations, any long-term assumption about human behaviour is nothing more than guesswork. The gradual perception that the Earth is limited, that its resources are limited and that the environment can only tolerate limited modifications will eventually create social and cultural feedbacks which are not predictable. These feedbacks will be adjusted to advances in technology that are still in the realm of science fiction. It is striking, in this connection, to contrast the imagination shown in science fiction with the lack of imagination about the probably more significant domain of evolution of human behaviour.

In the shorter term perhaps the most important fallacy in global demand projections that the foregoing discussion suggests is the use of a single world-wide average figure for consumption, despite the enormous disparity in living standards between the industrialized and underdeveloped countries. Conditions now occurring in

Calcutta would obviously be considered intolerable by the inhabitants of Paris or Boston, yet they are tolerated to a greater extent than is often realized. The degree of happiness does not seem as directly related to the GNP as the advertising industry would have us believe. In spite of the globalization of certain aspects of culture through the impact of mass media, fundamental differences in ways of life are not disappearing all that rapidly. In any case the hard facts of economic development are such that we are extremely far from anything like a uniform standard of living and to offer an average figure in these conditions is like quoting a mean for the body weights of an elephant and a mouse.

Two kinds of resources

Although the difficulties in assessing availability of resources may appear less formidable than the evaluation of needs, we are only beginning to appreciate the potentialities and limitations of the Earth's resources. The many alternatives offered by advancing technology can at any time change the frame of reference. We do perceive some absolute limitations, such as available space and the global heat balance, but these limitations are still too far ahead to give us clear guidance for today's courses of action, except to tell us to be more cautious and less wasteful.

Global figures for availability of resources are more meaningful than global figures for projected needs, when the resources in question can in fact be transported from one point of the globe to another. Such transportable resources, whether renewable or non-renewable, can be moved to where people need them. This applies to minerals, to liquid fuels and, to an ever greater extent, to power, water and food. This transportability is not altogether a blessing since most developing countries export raw materials for short-term gain and import food or finished goods – a procedure which hardly assists these countries in making economic progress. In evaluating the long-term availability of transportable resources, for global projections, some refinement is necessary to take account of the possibility that certain countries may withdraw certain resources from world markets, in order to use them themselves. In any case, transportability is a property of only one group of resources. Many other resources are non-transportable and many resource problems are directly linked to local environments. This is the case with natural amenities, with most water resources and obviously with climates, soils and wild genetic stocks. Future relationships between population and resources must therefore be considered largely in local or regional terms, since the world is not uniform and since we cannot imagine people moving away from their homelands in a world divided into nations. In other words, no matter how refined our global estimates may be, the resources problem still has to be faced at the local, national and regional level, for physical, biological and sociocultural reasons.

Global knowledge

We are thus confronted with an apparent contradiction between the need to look at the world as a whole and the need to look at its diverse local or regional sub-systems. Our knowledge of the latter, and of how they all somehow coexist and fit together is still very imperfect; even if we have a pretty good qualitative view of these local systems and their interactions, we obviously lack quantitative methods for their full analysis and the capacity to integrate and forecast their evolution. The contradiction can be resolved only by a global approach to studies of local systems: by vigorous and co-ordinated efforts to analyse the different environmental situations which exist in different parts of the world, from the tropics to the poles, and from the uninhabited forests to the old urban centres. We need to compare the functioning of these various sub-systems, and their responses to human intervention, and progressively to try to make a synthesis of all sub-systems which, together, constitute our living world.

The task is enormous but we are not starting from scratch. Much precursory work has been done, particularly in the past two decades, by geographers, climatologists, hydrologists, soil scientists, biologists, sociologists and others. Thanks to progress at the analytical level we understand a great deal about local phenomena and processes. As to comparison and synthesis, which obviously require international approaches, the productivity studies of the International Biological Program and the establishment of the Soil Map of the World (a joint venture of the Food and Agriculture Organization and Unesco) are examples of significant recent advances.

To arrive at more precise and quantitative conclusions for the rational management of environmental systems and their resources requires unprecedented co-operation in research at the international level. This is precisely the objective of the Man and the Biosphere programme recently launched under the auspices of Unesco. It is hoped that all countries, rich and poor, will work within this global research enterprise on problems of ecology and resources management which in one way or another affect them all. Other, more specialized, projects for global research and environmental monitoring are now being planned within the United Nations system, as a result of the 1972 Stockholm Conference on the Human Environment. Such global studies can use modern tools and pool the intellectual resources of all nations, for refining the analysis of natural systems and the complex effects of human action upon them as well as upon man himself.

They will be fruitless unless all nations improve their capacity to apply in practice the new knowledge about environments and resources. Better research institutions, management practice, laws and regulations are only a part of the requirement. Above all, people in every walk of life must be made aware of their partnership with nature and of the issues confronting them and their children. The hard choices which scarcity of resources imposes on mankind have to be explained before people can be expected to act in a responsible manner, to support necessary public measures and to adapt their personal behaviour to changing environmental conditions. Laissez-faire about population and resources is no longer possible and environmental education is necessary for every human inhabitant of our small and crowded planet.

References and further reading

Sterling Brubaker, *To Live on Earth* (Baltimore 1972).
Committee on Resources and Man, *Resources and Man* (San Francisco 1969).
R. F. Dasmann, *Environmental Conservation* (New York 1959, 1968 and 1972 and London, Sydney and Toronto).
P. R. Ehrlich and A. H. Ehrlich, *Population, Resources, Environment* (San Francisco 1969).
Food and Agriculture Organization, *Provisional Indicative World Plan for Agricultural Development: A Synthesis and Analysis of Factors Relevant to World, Regional and Agricultural Development*, I and II (Rome 1970).

W.W.Murdoch, *Environment, Resources, Pollution and Society* (Stamford, Connecticut, 1971).

SCEP (Study of Critical Environmental Problems), *Man's Impact on the Global Environment* (Cambridge, Mass. 1970).

Unesco, *Use and Conservation of the Biosphere*, Natural Resources Research Series, x (Paris 1970).

Unesco, *Scientific Framework of World Water Balance*, Technical papers in hydrology, 7 (Paris 1971).

Unesco, Final report of the International Co-ordinating Council of the Programme on Man and the Biosphere (MAB), First Session (mimeographed) (Paris 1972).

B. Ward and R. Dubos, *Only One Earth* (London and New York 1972).

LEARNING FROM
A DISASTER
L. H. N. Cooper

Living has always been dangerous. Primitive man had to find how to protect himself against unexpected storms and frosts or the attacks of predators who would make a meal of him if his defence failed. As civilization developed and man began to live in cities, new hazards developed, largely of his own making. One of these was the cataclysmic effect of uncontrolled fire in cities. Fire brigades developed, first as a band of neighbours and then as efficient, professional organizations run by public authority created for the purpose.

Technology, like living, is and will always be dangerous. Advanced communities have each and severally organized defences against the hazards which arise from new technology. There may be a lag of many years while news of solutions found by one community spreads to others having similar problems, and then becomes effective in action. Technology creates novel products and novel hazards. The appearance of a novel hazard usually creates an emergency to handle, for which there is little in the way of precedent. Authority has to extemporize as best it can with what it has.

Mankind learns only slowly from its mistakes and has not yet speeded up its ability to find out how to cope with novel technological hazards. Many fear that the accelerating rate at which new hazards are appearing is such that the price to be paid for new

L. H. N. Cooper was until recently deputy director of the Plymouth Laboratory of the Marine Biological Association. He is a chemist who has spent most of his career elucidating the physical and chemical environment of marine life. He is a Fellow of the Royal Society.

facilities and new amenities may be devastation of the world by an escalating number of ever-larger accidents.

The process of learning *ad hoc* will no longer suffice. A community has to use all necessary resources, not only vigorously to combat a novel emergency using men trained for action based on experience, but also to observe, record and reflect about the emergency in order to build up codified knowledge which can be applied when the emergency next repeats itself. Men of action are rarely effective at applying scientific method and all too often resent scientific observers seeking factual knowledge while contributing, so it seems, nothing towards conquering the immediate emergency.

The problem therefore is one of producing, as if out of a hat, a number of resourceful, tactful, trained observers who can note what happens during an emergency, can quickly design and carry through appropriate experiments and can draw conclusions as a guide to future action – all this without in any way impeding the work of the men on the job. Indeed, if mutual trust can be established, more immediate feedback may be possible. The observers, knowledgeable about ways in which similar emergencies have been met in other countries, with access to libraries and a few hours in which to think, may be able to suggest immediate improvements in operational methods.

It is important, though, that the two functions – reflection and operational action – should be kept apart. To attempt a mix is to court failure; to establish mutual trust and esteem between men with opposed temperaments is essential. Given these, it may be possible to hand over to the scientific observers a non-vital segment of the emergency situation to be used for experiments with built-in, designed checks and controls. From such experiments, which do not seek to simulate an emergency but are part of a real one, it may be possible to arrive at definitive answers, hard to get any other way.

Lessons from the Torrey Canyon *incident*

The *Torrey Canyon* incident may be used in illustration. On 18 March 1967 in clear daylight and in good weather this 970-foot tanker ran at about seventeen knots onto the Pollard Rock, one of a clearly charted group of rocks lying 25 kilometres west of England's Land's End and not far from some of the most attractive holiday beaches in the country. The tanker's cargo was 117,000 tons of Kuwait crude oil.

Her tanks ruptured. Escape of oil to the sea began at once and was to continue for several weeks, creating a coastal emergency without precedent. The Royal Navy began taking action within a few hours, on information supplied by the Royal Air Force. A control centre was set up at Devonport and the national government quickly assumed responsibility for marshalling all available resources for fighting oil pollution, which were considerable. The scale of the *Torrey Canyon* pollution and of the action needed to combat it had not been foreseen, but many contributed to build up an emergency operation which matched the need. There have been criticisms of the way the operation was conducted but these, relying on hindsight, are most unfair. The operators could work only with the received knowledge of their day and this they did most effectively for the duration of the emergency.

The known threats to the environment were two. The first was to the amenities for residents and holiday-makers by fouling of the beautiful sandy and rocky shores, in the first instance of Cornwall, then of the rest of Britain, of France and of any other countries of Europe which the oil might reach. A large part of the livelihood of the people of Cornwall and also of Devonshire depends on attracting holiday-makers so that powerful economic arguments were quickly made for keeping the crude oil off the beaches, at almost any cost. The chosen agents were 'detergents', better described as oil-dispersants.

The second known threat to the environment was the effect of the oil on all plants and animals, including sea-birds, living in the sea or on the seashore. Hardly anyone realized that there was a third threat, the vicious toxicity to marine life of the 'detergents' used to disperse the oil. The sharp lesson of the *Torrey Canyon* was that a remedy used to combat a novel environmental emergency may have an unforeseen backlash worse than the evil it is meant to cure. It fell to marine scientists not directly involved in the emergency action to chronicle the harm done by the 'detergents'.

At the time of the wreck none of us working at the nearby Plymouth Laboratory of the Marine Biological Association was concerned with the effects on marine organisms of noxious substances discharged into the sea. But we were naturally alarmed about the effect on 'our' sea areas and seashores of the oil leaking from the stricken tanker and of the agents that might be used to fight it. Our own researches were long-term and it was no light decision to interrupt them; nevertheless the *Torrey Canyon* disaster, on our own doorstep, was one which our laboratory was well-equipped to study and as citizens we had no choice. The Director led very surely but did not dictate and nine days after the stranding the staff in democratic assembly agreed to suspend all research programmes and to use every resource we had to study the consequences of the wreck. The Natural Environment Research Council offered the additional financial help that was needed.

By early next morning, our research vessel *Sarsia*, equipped for the unexpected, had left for the area of the wreck, parties of ecologists had started for the oil-polluted shores and a programme of laboratory tests of the toxicity of the oil and detergents had begun. Accommodation was found for other scientists who needed an advance base near the site of operations. For almost fifty staff and long-term visitors, personal disciplined involvement was complete and remained that way for ten weeks or more. Seldom can so much useful information have been obtained by so few scientists in such a short time; seldom can such a group of individualists have more willingly accepted direction and worked as a team. Within a year the results were collated and published as a book entitled *'Torrey Canyon' Pollution and Marine Life*. Besides highlighting the fatal effects of the 'detergents', especially when used for clearing beaches, the book dealt thoroughly with every aspect of the emergency as perceived by marine biologists.

Flexibility

Although specialist knowledge and methodology is essential, no laboratory is likely to have its skills precisely matched to emergency needs. It is therefore a source of strength if specialists whose particular skills are not needed can transfer their command of scientific method to complement the work of those on whom pressure is great. Thus, at Plymouth, one scientist with a gift for organizing data received all information about the *Torrey Canyon* pollution as it came in and not only filed it for future use but displayed it so that all his colleagues knew within a matter of hours what was going on. Two others made it their business to master this material and to man the telephone lines which were red hot with enquiries from the press and broadcasting and television stations, many of them from overseas. This was a most valuable service by professional scientists who passed on digested, accurate, balanced and readable accounts for publication. They also protected the field workers from being distracted from their primary tasks.

Clearly defined division of labour in an emergency is essential, but so is flexibility, as exemplified in our study of the drift of the oil. Observing oil at sea from the air is not easy and depends critically on lighting conditions. RAF Coastal Command, which made the observations, had no guidelines to help it plan its sorties as day followed day. Many sorties were wasted and there was a wishful belief that when the oil had drifted well away from the coast it would not come back. The aerial observations became very costly and the law of diminishing returns soon applied. When no more oil could be located the sorties had to stop, even though tens of thousands of tons of oil were unaccounted for and, as we now know, still represented a hazard.

We had early arranged with Maritime Headquarters at Devonport to provide us with copies of all factual observations which came in. Amongst these were the Coastal Command observations of the oil at sea, which our cartographer plotted each day. No pattern of events quickly hit my eye though, looking back, it should have done. The simple mechanism emerged only when our senior physiologist took the problem in hand. On a policy of 'each man to his own last' it was not his job but his freelance intervention led to one of our most important results – that oil on the surface of the sea is driven straight ahead of the wind at about 3.3 per cent of the wind speed. Such initiative and flexibility are essential if research laboratories are to learn as much as they can from emergencies.

The price of participation

The consequences for our laboratory were diverse, some good, some bad. We helped to fill an important gap in practical knowledge. Scientifically, pollution was introduced to us as a facet of our environment as important as is the weather, but no more so, and we have no wish to become narrowly pollution-oriented any more than we would wish to become temperature-oriented.

For myself, the subsequent five ill-organized years are not ones I would wish to live again. The turbulent aftermath of *Torrey Canyon* has been hard to shake off. Plans for rounding off a lifetime of carefully planned work on the physical and chemical basis of biological productivity in the sea have had to wait until after my retirement. There are probably many other similar debits which may be placed in the laboratory research account against the credits arising from our involvement in the *Torrey Canyon* affair.

A laboratory possessing the right skills should never shirk involvement in a grave emergency but, in my view, the price that has to be paid in loss of creative fundamental research is high. Long-term research programmes are fragile and only too easily wrecked by diversions to attack *ad hoc* problems thrown up by emergencies and for the scientists involved 'emergency action' is a loaded phrase. We are averse to cutting threads of work and thought which we may never be able to tie together again. Momentum lost during a ten-week crash programme may take years to recover. Emergencies are traumatic, not least for smoothly running research organizations whose primary duty it is to learn in every way they can from what happens in nature. Maybe, though, this personal view is biased; others with different ways of thought may disagree.

Preparing for the worst

Efficient governments now establish operational groups of resource-ful people primed with manuals epitomizing the experience of former emergencies. Such manuals are essential to achieve consistency and mutual understanding between people pressed into emergency service from many sources. Good manuals must be brief and easily read in a hurry. Much experience which could conceivably be needed sometime, somewhere, is better left out.

On the other hand, the amount of existing knowledge which may be relevant to an emergency may be vast, yet not accessible. Until recently very few environmental scientists had ever heard of methyl mercury or would have known where to look for information about it. Yet in 1971 the *Nordisk Hygienisk Tidskrift* was able to publish an expert group's report: 'Methyl mercury in fish: a toxicologic-epidemiologic evaluation of risks'. This book of 364 pages included about 530 references; it was very precisely and concisely written on a subject that was very new. Any group mounting emergency action in which mercury pollution was involved would need to have this book on hand in order to refer quickly to some of the quoted references.

As a young man I did not expect a librarian to do more than keep in order the books and periodicals I might need to do my job. I expected to do all my own reading and indexing. Latterly I have had to lean more and more on our librarian to help me with literature surveys, an admission which brings me no joy. At the time of the *Torrey Canyon* disaster the field and laboratory workers suddenly needed much information from disciplines alien to our normal work. Much of this our library service was able quickly to get for us but the strain on it was great.

Since then the pace of growth of new knowledge has accelerated still further. We now have a pollution information scientist whose task it is to keep track of all literature of the kind that will be needed when emergency action on marine environmental pollution next arises. Even this limited assignment is growing so much that the time of a single man is barely adequate to cope with the flow of new information. Thus highly trained, specialist information scientists are needed, who can quickly marshal from all sources pertinent knowledge in support of operational staff in an emergency. How to organize this service efficiently and economically needs much thought.

No one knows when or where the next environmental emergency will occur, by land or sea, or whether it will involve oil, poisonous chemicals, radioactivity or something we have not thought of. Government research establishments, such as the Warren Spring Laboratory and the fisheries laboratories in the United Kingdom, pursue relevant research and development work, often with experiments simulating accidents; but when an emergency arises they are stretched to the limit applying what is known, and the nation is lucky indeed if, in the meantime, something new has been learnt that helps in tackling the emergency. A scientist can undertake operational action and be equally successful at research but he cannot do both at the same time. So non-government scientists may still find themselves becoming involved.

A research institute or university laboratory, free from operational responsibilities and coming in afresh with basic disciplines, can do something which otherwise may not be done – observe, experiment and collate what happens during an emergency. The disruption of long-term research has to be tolerated, because only in this way can the lessons of each disaster be learnt and shared with other countries. It is essential that all countries learn from one another's experience, on how best to mount emergency action when it is needed, if the 'escalating number of ever-larger accidents' mentioned at the outset of this chapter is not to make the world a noisome place.

References and further reading

H. A. Cole (org.), 'A discussion on biological effects of pollution in the Sea', *Proceedings of the Royal Society* 177B (1971), pp. 275–468.

D. W. Hood (ed.), *Impingement of Man on the Oceans* (New York 1971).

Nordisk Hygienisk Tidskrift, 'Methyl mercury in fish', Supplementum 4, special issue (1971).

O. Schachter, *Report on Marine Pollution Problems and Remedies* (United Nations 1971).

J. E. Smith (ed.), *'Torrey Canyon' Pollution and Marine Life* (Cambridge 1968).

Teaching environmental science 26

TOWARDS A THIRD CULTURE
Leonard O. Myrup

The chief obstacle to dealing wisely with the environment, whether in research or in practice, is the mismatch in knowledge and attitudes among the many different kinds of experts involved. Difficulties of this sort have been discussed before, in other contexts, notably by C. P. Snow (1959), in his famous lecture on the 'two cultures'. Snow advanced the proposition that intellectual life in all Western society is now characterized by a pronounced division in fundamental viewpoint. This schism was supposed to be most pronounced between scientists and 'literary intellectuals'. Snow described both polar types as suffering from an over-specialized education that left them unable to understand or appreciate their opposites. In Snow's model, the scientists are thought of as hard-working specialists, uninterested in traditional high culture and, in the view of the literary intellectuals, filled with a shallow, perhaps dangerous, technological optimism. In contrast, the literary intellectuals are supposed to be anti-social individualists who are generally ignorant of the science and technology which is transforming the world around them. The existence of the 'two cultures', according to Snow, makes the problem of dealing with the

Leonard O. Myrup is chairman of the Division of Environmental Studies at the University of California, Davis, which was set up in 1970. He is a meteorologist by training, and uses computer modelling techniques to study relationships between geophysical, biological and social processes.

outstanding contemporary world problem, the gulf between the rich and poor nations, substantially more difficult to deal with. He argues that only a joint effort by the rich nations, using all intellectual resources, can possibly be effective.

The concept of 'two cultures' has been ferociously criticized by F.R. Leavis and others. In retrospect, it appears that Snow was primarily speaking of British society and that his analysis has lesser relevance in other areas. For instance, his description of the leaders of art and literature as politically reactionary and socially irresponsible does not seem applicable even in other Western countries, let alone the rest of the world. Snow's central phenomenon is nevertheless real and demonstrable: the existence of large differences in education, values and fundamental attitudes between the various groups of educated persons. For instance, E.C. Ladd and S.M. Lipset (1972) have recently studied the political attitudes of some sixty thousand faculty members of American universities and colleges. When responses are organized into an overall index which defines a liberalism/conservatism scale, remarkable contrasts are found between the various academic disciplines. The most conservative groups were agriculture, engineering and business faculties. The most liberal in political orientation were faculties from social science, the humanities and law. It is a striking and perhaps ironic fact that solution of many of the most pressing world problems would require input from most if not all of these very disciplines.

Of course, political orientation is only a small part of professional life. There is some evidence, however, that more general attitudes and even personality indices are correlated amongst members of given disciplines. For instance, M.C. Regan (1966) has found pronounced similarities in personality structure among members of given university faculties, and also among their students. It is not known how such groupings of human attitudes occur. In professions it is probably the result of strong association with a specific clientele in society reinforced by common training, methodology, sources of information, literature and professional associations. Whatever the origin, it appears that there are measurable differences between disciplines and professions that may act to impede exchange of information and co-operative problem-solving.

In retrospect Snow's emphasis on industrial development in the poor nations seems even more timely now than in 1959. The confrontation which occurred between the Western environmentalists and representatives from the developing nations at the United Nations Conference on the Human Environment at Stockholm in 1972 showed that the priorities for the less affluent portions of the world remain solidly with industrial development. A similar viewpoint has appeared among American leaders of minorities, labour unions and inhabitants of the inner cities. Jobs, housing and health care are seen as much more important than the rich man's concern with 'preserving the quality of the environment'.

Thus, the picture which is emerging from our experience with the recent preoccupation with environmental problems in Western countries is complex. On the one hand, in the developed countries important resources are being directed to the effort to preserve or restore environmental quality. New legal regulations, technological and educational institutions are arising which promise effective responses to the problems of the environment. On the other hand, the age-old dilemma described by Garrett Hardin in his classic essay 'The Tragedy of the Commons' (1968) has not been resolved because the 'common-land' that we are overtaxing is most certainly global in extent. For instance, the use of DDT in the United States has almost ended but huge amounts are still being sold to the developing countries. Considering the monumental benefits, on the short term, for human health and food production, few would advocate ending world DDT use at this time. Similarly, world energy demand is rising at such an incredible rate that it appears inevitable that the Alaskan North Slope and all other known oil reserves will be developed within our lifetime. The poor nations will develop an industrial base if they possibly can and poverty in the affluent nations must be dealt with. It seems that economic and social development remains in fundamental conflict with the need to maintain environmental quality.

I suggest that the diverse set of issues and problems touched upon so far have important common factors. The gulf between Snow's two cultures is paralleled by the division in viewpoint between environmentalists and the advocates of economic and social development. However, closer inspection of this particular dilemma shows that both aims must be attained or neither will be possible. Historically, pollution, resource depletion, urban crowding, poverty and human misery have been closely associated with economic development. While social and industrial development in third-world countries is of the highest priority, the social cost, in the long run, of ignoring the environment will be disastrous. We must develop a new consciousness which can balance social, economic, techno-logical and environmental needs and processes in planning for the next century.

The educational challenge

The challenge to educationalists is to create a learning framework for students of the environment which is broad enough to include the social, economic and technological aspects of environmental programmes as well as their deeper links with fundamental human values and life styles. I propose that a broad education which meets this challenge is also the road to the creation of a 'third culture' which bridges the gap between Snow's 'two'.

In many ways the students are ahead of us. Well before the ecology boom of 1970, student groups abounded whose objectives and tactics were directed towards preserving the environment in a highly energetic, pragmatic and non-revolutionary manner. These students are still with us in greater numbers than ever and they move toward their objectives with great elan and verve. Their approach to human problems of the environment is intelligent, problem-centred and takes full advantage of opportunities to work within the existing social system. They realize, as one of my col-leagues has put it, that 'a lot of the barriers to going out and solving problems are totally non-existent because you don't have to be an expert to go out and solve problems. What you have to be is an inquiring person who looks for relationships and tries to be useful'.

The social and environmental problems in society are difficult for the universities to respond to. The classical organization in uni-versity teaching and research has been formally to divide problems into smaller and more manageable 'disciplines', which are treated in great depth. This approach has been extraordinarily successful in advancing some of the basic sciences, but has been less successful in training students for non-research careers. The disciplinary department has been almost totally incapable of dealing with broad, real-world problems either in teaching or research. The barriers between the departments which compound Snow's 'two-cultures' model effectively hinder cross-disciplinary work.

The educational objective in environmental science should be to create a new generation of 'third-culture' professionals. They should understand and appreciate the tools and capabilities of most of the technical specialities whose input is essential to the problem of maintaining and improving the quality of human life on this planet.

Emergent patterns

The period since 1970 has seen a rapid and diverse development of environmental teaching programmes in the United States. Usually named Environmental Sciences or, more commonly, Environmental Studies, they attempt a broad, interdisciplinary synthesis of those areas of human knowledge relevant to environmental problems. Strong emphasis is always placed on the biological and physical sciences. The prominence of the social sciences, engineering and the humanities, on the other hand, varies widely between institutions but the balance between the various fields is only one of the distinguishing polarities of the different approaches adopted. Others are flexibility and breadth versus rigour and depth, basic science versus problem-solving, university versus community orientation and research versus teaching. In order to be specific I shall discuss particularly the programmes which have developed on the various campuses of the University of California.

The most common pattern which has emerged is of a major programme, administered by an interdepartmental committee, usually centred in the College of Letters and Sciences. The curriculum, which is moderately structured, includes introductory courses in the physical, biological and social sciences and a small number of specially designed core courses. The student is normally expected to choose one area for emphasis which may be, but usually is not, highly specialized. Senior undergraduates often participate in a problem-specific seminar course which treats a real-world problem in depth. Instructors remain primarily associated with their disciplinary research departments and teaching assignments tend to be voluntary labours of love and often temporary.

These programmes have the inherent disadvantage that committees are weak relative to research departments, when competing for university resources. Most faculty members are reluctant to participate in an interdepartmental programme if this simply means a heavier teaching load. Even if the university does allocate funds to purchase faculty time, chairmen of departments are by no means eager to co-operate, because of the practical difficulty of finding suitable temporary instructors to replace those 'lost' to environmental studies. One suspects that the committee system is only an interim solution and that long-term survival will depend on developing a stronger institutional framework.

Such weaknesses, though, should be considered in their context. After several decades of unparalleled growth, the University of California has been forced to operate within a limited budget in recent years. Consequently, the development of new programmes, particularly if substantial new resources are involved, is not easy. Even so, high priority has been placed on environmental problems by the university community and several teaching programmes have emerged.

The environmental studies programmes on the Santa Barbara and Berkeley campuses of the University of California are examples of interdepartmental, committee-administered teaching programmes. The Santa Barbara programme features a set of lower division introductory courses equally divided between the biological, physical and social sciences. Official requirements for the major include sufficient concentration in a departmental course to enable students to qualify for entrance to graduate school.

The Environmental Studies major on the Berkeley campus, as at Santa Barbara, is administered by the College of Letters and Sciences and is organized into three areas of specialization in the physical, biological and social sciences. Students in all three areas receive training in the fundamentals of physics, mathematics, chemistry and biology. The contribution from the social sciences is noticeably less developed than that of the natural sciences. Even students opting for the social-science area are required to take eleven courses in mathematics and the natural sciences but only two social-science courses during their first two years, while students in the natural-science areas need take no social science at all. Second-class citizenship for the social sciences is unfortunately a characteristic of most attempts to develop interdisciplinary teaching programmes in environmental studies.

The Conservation of Natural Resources (CNR) curriculum on the Berkeley campus is quite different. The College of Agricultural Sciences and the School of Forestry and Conservation administer it jointly and describe it as an 'experimental field major'. The undergraduate course is loosely structured, extremely flexible and

strongly problem-oriented. Like the Environmental Studies programmes described above, this one is directed by a committee, but
with very substantial student participation. For instance, students
can initiate courses and write proposals for support funds. The
system for advising the student emphasizes flexibility, allowing him
to locate a professor with whom he can communicate. The major is
organized around three core courses, which depend heavily on
large numbers of guest lecturers. In his later years the student has to
specify 'area of interest', which may be a particular job he has in
mind, a discipline or problem that interests him, or preparation for
graduate school. One of the most striking aspects of CNR is the
strong, even dominating roles played by undergraduate and graduate students and the junior faculty. The senior faculty appear to
play no role except as invited guest lecturers. All of the core courses
stress the problem-solving experience in the community and students have participated effectively in the local decision-making
processes regarding such diverse phenomena as shopping centres
and neighbourhood mini-parks. In contrast with the more formal
Environmental Studies programme on the same campus, large
numbers of students participate in CNR. In 1972 enrolment in the
introductory core course was almost 500 and there were more than
150 students enrolled as CNR majors.

In neither CNR at Berkeley nor Environmental Studies at
Berkeley or Santa Barbara does faculty research play any significant
role. The members of the various administrating committees
typically continue their specialized research under the auspices of
their 'home' department. Of course, many of their individual
research programmes are highly relevant to the environment.
However, such activities do not usually constitute the sustained,
co-ordinated research which would be an effective response to the
environmental challenge. In keeping with our earlier remarks, what
is needed is research appropriate to the 'third culture'; that is to say,
problem-centred, truly interdisciplinary programmes with long-
term objectives closely related to the solution of pressing problems
of man and the environment.

The Environmental Studies programme at the Davis campus of
the University of California contrasts with those discussed so far.
Recognizing the weakness of committee-run systems and the
importance of interdisciplinary applied and basic research, we have
'institutionalized' environmental studies so that long-term goals in

teaching and research might be achieved. At Davis, an inter-college unit called the Division of Environmental Studies (DES) was formed, with the assistance of grants from the Rockefeller Foundation. Substantial university resources were allocated to DES including twelve full-time faculty positions, appropriate research facilities and, in 1973, a new building. DES faculty members are freed from the liabilities of full-time membership in disciplinary research departments.

Our full-time appointments span the areas of systems ecology, community ecology, limnology, animal behaviour, political science, systems engineering, biometeorology and physics; this constitutes the 'core faculty' in the Division. In addition, part-time or split appointments have been made in such areas as sociology, political science, agronomy and engineering. As a result, strong research projects are developing in, for example, the management of natural ecosystems, land-use planning and an analysis of governmental policy towards environmental problems. Graduate students play an important part in the research programme, which is designed to provide the central thrust for environmental research on the Davis campus.

The processes by which university teaching and research are carried through to implementation in the community are an essential part of the DES programme. 'Information delivery' to and from the community is facilitated by environmental specialists with specific duties in this respect. Their formal teaching assignments are reduced or eliminated and they are able to organize a variety of programmes for which faculty members with normal teaching obligations would have insufficient time. For example, various short courses, conferences and workshops have been developed around the application of ecological principles in the management, decision-making or planning processes in society; participants have typically been members of state and federal regulatory agencies. The specialists in question also play a principal role in the DES 'intern' programme, in which students go to work with governmental agencies, with committees of the state legislature or with environmental organizations in the private sector. On the whole, the institutional framework, research and the extension process to the community are better developed on the Davis campus than for environmental studies at other University of California campuses.

The concept of undergraduate instruction in environmental studies also differs at Davis. Rather than administering a major programme of its own, the Division of Environmental Studies has aimed at developing an environmental consciousness among undergraduates in all areas. In co-operation with various departments, courses are offered which relate environmental problems to various areas of knowledge. These courses are usually taught by a team, with a minimum of reliance on guest lecturers. They emphasize the systems approach to the environment, techniques of mathematical modelling, and the relevance of the basic biological, physical and social sciences. Even if a major programme is developed in the future at Davis, we shall continue to regard our primary teaching task as being the infusion of the environmental ethic into traditional courses, and making use of existing resources rather than generating competing demands for scarce faculty, student, and research facilities.

Looking ahead

The future of environmental science in universities in the United States and elsewhere depends on three factors. The first is moral and strategic. The environmental movement must develop in close co-operation with the efforts of the developing nations and of the poor and hungry sectors of affluent nations to achieve economic and social equality. The pressing needs for food, housing, medical care, jobs and energy are real and demand the highest short-term priority. If the situation develops into a confrontation with the legitimate aspirations of the poor and the hungry, then the environment will lose again. We must work towards that 'third culture' which recognizes both the importance of rapidly satisfying human needs and, at the same time, the long-term goal of maintaining or improving the quality of the environment.

The second factor is administrative. Environmental concerns must be institutionalized at all levels of society, of which the universities are only one. If short-term enthusiasm and voluntary associations are to be replaced by effective public policy, legislation and regulatory procedures, an informed and interested public is one prerequisite; another is a new generation of 'third-culture' administrators and managers, broadly trained and able to see both human and environmental needs in perspective. In this context, the

275

universities need to institutionalize their present environmental teaching so as to make a permanent commitment of faculty and resources. Too many of the present programmes are built on shifting academic sands. The inclusion of the social sciences and the development of problem-oriented long-term research are particularly critical to this process of institutionalization.

Thirdly, the universities must produce practical payoffs.Environmentalists have done a superb job in alerting the public to the basic and unarguable facts of the environmental crisis, so that important human resources have been diverted towards finding solutions, particularly in the affluent countries. This effort must show results or we shall have lost a chance that we may not get again. The processes by which the results of university research and teaching are extended into application and implementation must be taken seriously. Publishing results of research in the academic literature does not, by itself, solve any problem. University faculties should be willing to work with legislative bodies, regulatory agencies and with community groups. Hopefully, a new conception of the professor will arise which encourages his participation in society as well as in the traditional functions of the university.

References and further reading

G.Hardin, 'The tragedy of the commons', *Science*, 162 (1968) pp. 1,243–8.

E.C.Ladd and S.M.Lipset, 'Politics of academic natural scientists and engineers', *Science*, 176 (1972), pp. 1091–100.

F.R.Leavis, *'Two Cultures? The Significance of C.P.Snow* (London 1962).

M.C.Regan, 'Relationship of selected personality variables to college students' elected fields of major study', paper presented to California Educational Research Association, Stanford University (1966).

C.P.Snow, *The Two Cultures and the Scientific Revolution* (Cambridge 1959).

ET CETERA
Nigel Calder

The mental task to which this book represents one approach is nothing less than to find new, comprehensive ways of thought. The scientific strategy of isolation and dissection, which has served so well for three hundred years, may be as inappropriate for the wise management of the environment as exhaustive examination of a single note would be to the composition of a melody. I say only that traditional methods of research *may* be inappropriate, because what is offered for putting in their place is by no means well proven.

The motive in producing this book was dismay at the intellectual sloppiness of many utterances about the environment and at the way ecology came to be regarded as a political movement rather than a science. I have satisfied myself, at least, that there exists in 'ecology plus' a broader and deeper understanding of environmental issues than we are normally enabled to see.

There has been no sense of scraping a barrel. On the contrary many approaches and examples in environmental science have been left out, in keeping the book to a convenient size. One fascinating area of omission concerns human adaptability under environmental stress, the influence of heredity and the knotty issues of natural selection and ongoing human evolution. Another, of a quite different kind, is the methodology worked out by the international Scientific

Nigel Calder is a science writer. Earlier books touching on environmental science include *The World in 1984* (edited), *The Environment Game*, *Technopolis* and *Restless Earth*. He was editor of *New Scientist* until 1966 and since then has scripted several major television programmes.

Committee on Problems of the Environment for global environmental monitoring using 'baseline' stations far removed from sources of pollution, to measure changes of climate and worldwide pollution by heavy metals, DDT and other materials. High-flying aircraft and satellites equipped with cameras and other remote sensors are providing a novel technology for environmental surveys. Many natural phenomena, the hidden cosmology of the soil for example, have been neglected in this book. On the other hand, the social sciences are certainly less fully represented than they might be; a logical discontinuity seems to exist at present between insect behaviour (for example) and the processes studied by political scientists and lawyers.

Two aspects of the future of environmental science receive brief comment in this concluding chapter. What are the prospects for sharpening the methodologies? Looking beyond the immediate battles about pollution and resources, what future uses can one envisage for successful environmental science?

'The science of everything'

The difficulties that teaching institutions encounter in trying to deal with environmental issues run much deeper than a temporary mismatch between traditional academic departments and present social needs. The dilemma of specialization versus breadth in human thinking is a real and ancient one. For discovery and invention in environmental science, specialist dedication seems indispensable and there is growing experience in recruiting and using specialists in interdisciplinary projects. Although environmental science is 'applied' research, it cannot be unconcerned with advances in knowledge at a fundamental level. Just as the needs of medicine have set the pace in many areas in discovery, from microbiology to brain mechanisms, so environmental science will encourage attention to the interactions of genes and environment, to the geological history of the atmosphere and to many other fundamental questions.

Will there be a role for general environmental scientists, serving the same sort of function as the family doctor does in medical care? A patient with influenza does not go to a molecular biology laboratory for treatment and a mayor would not take his waste-disposal problems to a global computation centre. Some government scientists in Washington and London think that national, regional and city governments, and industry as well, will indeed want environmental generalists – 'ecosystems managers'; others emphasize, instead, the need for competent systems analysts almost irrespective of their environmental knowledge. A third view is that the one-man environmental practitioner could not have wide enough knowledge but that 'group practices' may emerge in the form of multidisciplinary teams able to respond to most environmental issues put to them. What professional way of life will develop depends on actual career opportunities in government and industry and on the success of independent consultancy groups. The greatest scope for the environmental generalist may be in the developing countries, where administrative hierarchies have not yet ossified.

Meanwhile, many trained specialists need to become semi-generalists by learning about a spectrum of other specialists' approaches to the environment, as exemplified in this book. On a considerably larger scale there ought to be a demand for technicians able to trap insects or analyse water or carry out other tasks in monitoring the environment – unless we are going to have more chiefs than Indians and build grandiose models on scanty data.

These practical uncertainties about manpower illustrate the fact that the invention of environmental science has not proceeded very far. The most pressing task is to make the subject-matter and the methodologies more manageable.

Words of Voltaire ring uncomfortingly: 'Madness is to think of several things too fast, or of one thing too intensively'. The latter part of his aphorism is now well accepted; in environmental science the watchwords are 'holistic', 'multifactorial', 'interdisciplinary'. The obsessions that allowed engineers to ignore pollution and farmers to regard all uncultivated species as pests or weeds seem definitely to be remitting. But what of the schizoid thoughts that may replace them – 'several things too fast'? The present symptoms are of two contrasting kinds: the neurotic wave of the arm as the would-be analyst indicates ('et cetera, et cetera . . .') a host of interactions that he means to get to grips with some day but not now; and the psychotic smirk of the man who has already weighed mankind in a computer and found it wanting.

We should neither take fright at the complexity of the environment nor be so rash as to want to come up with 'the science of everything' overnight. Even if, old hunters all, we are smart enough to grasp many intricacies of ecosystems, our means of communication may cramp our style. You only have to look at library classification systems, or recall the failure to get computers to translate human languages adequately, to see that we can externalize only in approximation the potentialities of human thought. Beyond these difficulties of articulation it is still an open question whether nature in the round can ever be, in principle, as comprehensible as atomic nature has happily proved to be.

Another parallel can be drawn, this time within environmental science itself. Meteorologists simply do not know whether the behaviour of the atmosphere is inherently determinate or capricious. They are nowadays using the most powerful computers that exist in an effort to find out. That does not stop them doing a useful job meantime and the fact that their predictions even for tomorrow's weather are sometimes wrong is a salutary reminder of the mysteries of the environment and a spur to research.

The pragmatic principle implicit throughout this book is that if you start by looking closely at nature you are being scientific and if you do this to some practical end it may be useful. That the bedrock of environmental science is close observation of the environment seems not to be as self-evident as it ought to be, nor does everyone who offers prescriptions for the environment take the trouble to look out of the window. As Smith remarks:

> The lesson to be learnt by environmental scientists from those experienced in applied meteorology is that every environment is different and inconstant, and that complex systems can be understood only by meticulous data-collection, logical analysis and repeated practical-scale investigation to identify the governing factors in each real situation. Neglect of any part of this combination leads quickly to nonsense.

To these prescriptions we can add the scientist's happiest mode of working, namely experiment. As Schindler and Huffaker make clear, experiments with 'life-size' ecosystems are both possible and rewarding. Academic ecologists, too, have made important experiments on the number of species an island or enclave of a given size can support. And of course environmental science cannot afford to be so fastidious as to ignore what can be discovered from uncontrolled perturbations of the environment which agriculture and industry set up every day.

Description, perturbation, logical synthesis, prediction and rechecking by description – this is the formula for advance in environmental science. The rate of advance will depend on the wit of the individual scientist but also on the manpower and other resources available. For example, there is nothing in environmental biology or pollution control that yet begins to compare with the meteorologists' globe-encircling network of observing stations, ships, weather satellites and balloons, and their computerized world data centres. There is some danger that governments will go on demanding far-reaching environmental advice without creating the essential organizations – like asking a lone man with a barometer for a long-range weather forecast.

The computer is to environmental science as the knife is to anatomy. While the knife takes things apart the computer puts them together and it is the chief aid to solving Voltaire's problem of 'several things too fast'. Wilson, French and Watt describe a clear target for environmental science in ever more comprehensive computable models of environmental systems. These contributors differ in their judgements of present accomplishments – including the *Limits to Growth* study that was the subject of much publicity, praise and disparagement in 1972. The issue is really about the limits to computability.

Watt in particular argues that a systems model, being concerned with relationships rather than with absolute values, can transcend the many deficiencies or uncertainties in the assumptions and data used. Some critics insist on the adage of the computer age: 'garbage in – garbage out'. A compromise is to say that one reason for making models is precisely to expose present weaknesses in data.

Although it is a fallacy that computers can do nothing but arithmetic, their use in modelling introduces a bias towards numerical data and there is always the risk that non-computable problems will come to be regarded as non-problems. To take a conspicuous example, relevant to global forecasting models, no one can quantify the impact of an invention that has not yet been made, but more qualitative discussions can take such possibilities into account.

Even in the simplest models, at the most elementary arithmetical level, tricky questions arise about whether quantities should be summed or compounded. Do you, for instance, simply add together deaths 'due' to disease and deaths 'due' to malnutrition or do you, at some risk of obfuscation, try to express the greater vulnerability of the hungry organism to disease? The answer depends on whether the model is more concerned with total population or with nutrition. And there, in the end, is the way in which 'the science of everything' preserves its mental health: by addressing specific practical questions, ranking terms and factors that might bear upon it, and deciding which can, for the moment, be neglected. The et ceteras have a cut-off point and the goal is to identify the governing factors in a particular situation.

Put at its lowest, environmental science provides data on the environment, a check-list of possible interactions ('don't forget the microbes!') and insight, from particular cases, into the forms these interactions may take. Beyond these rudiments, exciting possibilities are opening up for developing models. Just as the physical sciences have prospered by observation and experiment reacting endlessly with theory and speculation, so the way ahead in environmental science is clear enough. There must be imaginative interplay between the man at the computer keyboard and the man in muddy boots, with neither supposing that he can do the job alone.

Radical possibilities

Environmental science emerges to meet the needs of a world undergoing rapid change and it would be futile to consider the future of environmental science without taking account of other trends in human affairs. For a start, even if developing countries prove to be more modest in their material ambitions than Americans and Europeans, the already unavoidable growth in population will require far-reaching environmental transformations to feed, clothe, house and employ the newcomers.

For better or worse, technology dedicated to pressing human needs will continue to set the pace of really radical changes in the global environment. Hard on the heels of the Green Revolution will come another product of bio-technology: the Nitrogen Revolution. The release of a common soil bacterium re-equipped by man to fix nitrogen from the air could greatly increase the 'natural' supply of nutrients to the soil to the point where farmers would no longer need artificial nitrogenous fertilizers. This is not science fiction but the subject of promising experiments, notably those sponsored by Britain's Agricultural Research Council at the University of Sussex. The impact on the environment is literally incalculable at present; inevitably there will be anxieties about the microbe in question proliferating out of control, altering the rules of life in all natural systems on land and in our rivers and lakes.

By such alterations in basic components in environmental systems man has the power to transform this planet profoundly. The practical consequences of the far-reaching discoveries of the molecular biologists in the past two decades have still to appear, but it is only a matter of time before man will be able to 'invent' new species almost at will. Meanwhile, the more conventional approaches of the plant breeder are being applied to making cereals in which plant protein, poorly matched to human dietary needs, is altered in quality to something more like animal protein. This innovation could greatly improve human diet while reducing the loss of free energy involved (as Nobel explains) in transforming the protein by way of real animals.

All sorts of opportunities exist for reconstructing our food species, domesticating new species and making food by unconventional or artificial means. They are limited more by dietary conservatism and by the available energy than by technical hurdles. Those who are anxious about the increasing pressures of agriculture upon wildlife should be demanding the development of whatever of these techniques are needed to make sure that man feeds his growing armies of surviving children without enlarging the area of land under the plough and the hoof.

Better still, new ways of growing food should eventually diminish the territorial demands of the farmer, who has long since appropriated the choicest areas of the biosphere. If so, there will be major opportunities in 'wildlife engineering' – not merely letting ground go fallow but trying by careful reassembly of species to reconstruct ecosystems like those existing before man started farming. We could go further, and use our engineering skills to create a variety of novel wildernesses in which processes of natural evolution, distorted by man's activities, can resume with new vigour and eventually increase the diversity of living species.

Nothing will be gained by rescuing land from the farmers if it is taken over by the builders of houses and roads. The overall aim of urban planning should be to alleviate present evils of congestion and overcrowding, by better transport systems and more intelligent arrangements of living, working and recreation spaces, but without increasing the total area occupied by urban and suburban man. Again, diminution of the areas would be better.

The 'change of heart' which has made concern about pollution and the quality of life a force to be reckoned with in economic development helps to reinforce a trend that was already incipient in the rich countries. Patterns of consumption are likely to ease away from endlessly increasing demands on energy and materials towards pursuits such as tourism, electronic entertainment, extended education and spaceflight which are much more sparing of resources than, say, the automobile industry. To take one of these, tourism, its economic importance in encouraging conservation of cities, countryside, beaches and wildlife should not be underestimated, even though visitors leave minor environmental problems in their wake.

One need not share Singer's unmitigated enthusiasm for nuclear energy to see that fossil-fuel machinery, including the private use of the internal-combustion engine for transport, is eliminating itself by using up the economic reserves of buried carbon. This tendency, together with the increasing diversion of human effort towards working with information rather than with metals, will alter the character of the human impact on the environment, and afford scope for reorganizing economic and social systems towards environmentally benign forms.

It is not too early for environmental scientists to abandon their defensive, conservationist posture in favour of something much bolder and more constructive. In a rapidly changing world they should be in the thick of the action, proposing rather than opposing, suggesting how shrewd use of environmental processes can enrich our lives. So much rural and urban development is going to happen in the next thirty years that there is no reason, except lack of imagination, why we should not involve millions of human beings in hundreds of full-scale experiments, letting them act out, in their lives, man's search for models of renewed harmony with his environment.

Like others who see interesting environmental possibilities in current social and technological trends, I am well accustomed to accusations of complacency or (even worse!) of optimism, from that strange eco-alliance of misanthropists and Maoists, neo-

Luddites and health-food merchants. The revolution that I have in mind, though, is much more far-reaching than theirs. For myself, I should not be content unless England looked again as it did to Clark's stone-age aviator: 'great tracts of forest, broken only by the upper ranges of mountain chains, by forest glades, and by rivers, lakes, marshes and coastal lagoons.'

References and further reading

N. Calder, *The Environment Game* (London 1969 and as *Eden Was No Garden* New York 1967).

NASA (National Aeronautics and Space Administration), *Ecological Surveys from Space* (Washington,DC 1970).

National Science Board, *Environmental Science: Challenge for the Seventies* (Washington,DC 1971).

SCOPE (Scientific Committee on Problems of the Environment), *Global Environmental Monitoring* (Stockholm 1971).

Unesco, *Interactions between Environmental Transformation and Genetic and Demographic Changes*, MAB 4 (Paris 1972).

Index

diversity, 230-1
Dobe, 64-5
Doderer, 181
Doutt, R., 183, 189
Doxiadis, C. A., 91, 94
Dust Bowl, 206

earth sciences, 25-35
earthquakes, 30
East African Agriculture and
    Forestry Research Organiza-
    tion, 170
ecology, 18-24, 48-67, 120-1,
    175-81, 183-91, 203-13
economics, 119-27
economic growth, see growth
ecosystem, definition, 18-19
Ecumenopolis, 91
education in environmental
    science, 21, 267-76
Einstein, A., 161
Ekistics, 91-100
electricity, 80, 111, 240-1
Elton, C., 18, 190
energy, 25-7, 37, 52-3, 157-67,
    236-45; demand for, 111,
    217-19, 236-46, 269; see also
    free energy
engineering, 30, 110-18, 193
ethylene, 54
eutrophication, 176, 181
evaporation, 10, 12, 39, 170-3,
    212, 253
Experimental Lakes Area, 176

feedback loops, 217-20; negative
    feedback, 146, 154, 186; posi-
    tive feedback, 153, 217
fertilizers, 54, 178, 228-30
fish, 5-16, 62, 233, 251

Fisheries Extension Service, 13
Fisheries Research Board, 176
floods, 174
food, 61-7, 250-1
food additives, 102
Food and Agriculture Organiza-
    tion (FAO), 256
food chains, 166
food poisoning, 102, 104-7
food sharing, 63, 65
Ford Foundation, 228
forest, 48, 69-77, 180-1, 184,
    226, 284; see also deforestation
Forrester, J., 221
fossils, 69, 143
fossil fuels, 26-8, 32, 149-51,
    223, 236-40; see also coal, oil,
    gas
Fox, Sir C., 74
Foyn, S., 194
Fredriksen, R. L., 181
free energy, 157-67
Fremlin, J., 155
French, N. R., 281
Fries, M., 69
frost, 42-3
fungus, 190
furnaces, pollution by, 238-9
futurology, 247

garbage, see waste disposal
gas, natural, 26, 223, 236-7, 252
gasoline, 236, 238-9
geology, 25-34
geothermal power, 240-1
Gibbs free energy, 158
Giffen, A. E. van, 73
global regulation, 143-55
glucose, 164-5
Godwin, Sir H., 69

Kaelas, L., 77
Kalahari Desert, 63
Kenya, 59, 169–71
King William Island, 63
kōgai, 82
krypton, 243

Ladd, E. C., 268
Lake 227, 176–80
Lake Chilwa Co-ordinated Research Project, 5–16
landscape, 75–6
land tenure, 14–15
lead, 238, 252
leafhopper, 189, 229
Lean, D., 180
leaves, 52–3, 55
Leavis, F. R., 268
Lee, R., 63–4
legume, 50–3
leprosy, 8
Liebman, J., 116
Lieth, H., 207
lighthouse analogy, 123
light saturation, 53
Likens, G., 177, 180
Lindeman, R., 203
lithium, 244
Lipset, S. M., 268
Littleton Bog, 70
locusts, 8

maize, 8, 53, 185, 223
malaria, 8, 22, 107
Malawi, 5–16; Univ. of, 7–8, 10–11
Malaysia, 233
Man and the Biosphere, 257
Marine Biological Association, 261

Marks, D. H., 116
Marshall, L., 64
Massachusetts Institute of Technology (MIT), 137, 149, 223
mathematical models, *see* systems analysis
Mbarali, 173
McCulloch, J. S. G., 170
Meadows, D. H. and D. L., 137, 223
mealybug, 212
Mech, L. D., 188
medical science, 101–8
Mediterranean, 99
megalopolis, 94
mercury, 252; methyl mercury, 82, 264
metal ores, 28, 252
meteorology, 25–35, 280; *see also* climate, weather
Mexico, 228
micro-organisms, 54, 103; *see also* bacteria
millet, 8
Milner, J., 107
Minamata disease, 82
minerals, 28, 251–3
Mitchell, F., 70
monitoring, 38–9
Monks Wood Experimental Station, 19
models, *see* systems analysis
mongongo nut, 64
Morgenstern, O., 124
Morris, R., 124, 184
mosquitoes, 107
motor vehicles, *see* automobiles
Mount Fuji, 81–2
Mullen, G., 153